En vertu de l'article 1er de la Convention signée le 14 décembre 1960, à Paris, et entrée en vigueur le 30 septembre 1961, l'Organisation de Coopération et de Développement Économiques (OCDE) a pour objectif de promouvoir des politiques visant :

- à réaliser la plus forte expansion de l'économie et de l'emploi et une progression du niveau de vie dans les pays Membres, tout en maintenant la stabilité financière, et à contribuer ainsi au développement de l'économie mondiale ;
- à contribuer à une saine expansion économique dans les pays Membres, ainsi que non membres, en voie de développement économique ;
- à contribuer à l'expansion du commerce mondial sur une base multilatérale et non discriminatoire conformément aux obligations internationales.

Les signataires de la Convention relative à l'OCDE sont : la République Fédérale d'Allemagne, l'Autriche, la Belgique, le Canada, le Danemark, l'Espagne, les Etats-Unis, la France, la Grèce, l'Irlande, l'Islande, l'Italie, le Luxembourg, la Norvège, les Pays-Bas, le Portugal, le Royaume-Uni, la Suède, la Suisse et la Turquie. Les pays suivants ont adhéré ultérieurement à cette Convention (les dates sont celles du dépôt des instruments d'adhésion) : le Japon (28 avril 1964), la Finlande (28 janvier 1969), l'Australie (7 juin 1971) et la Nouvelle-Zélande (29 mai 1973).

La République socialiste fédérative de Yougoslavie prend part à certains travaux de l'OCDE (accord du 28 octobre 1961).

L'Agence de l'OCDE pour l'Énergie Nucléaire (AEN) a été créée le 20 avril 1972, en remplacement de l'Agence Européenne pour l'Énergie Nucléaire de l'OCDE (ENEA) lors de l'adhésion du Japon à titre de Membre de plein exercice.

L'AEN groupe désormais tous les pays Membres européens de l'OCDE ainsi que l'Australie, le Canada, les États-Unis et le Japon. La Commission des Communautés Européennes participe à ses travaux.

L'AEN a pour principaux objectifs de promouvoir, entre les gouvernements qui en sont Membres, la coopération dans le domaine de la sécurité et de la réglementation nucléaires, ainsi que l'évaluation de la contribution de l'énergie nucléaire au progrès économique.

Pour atteindre ces objectifs, l'AEN :

- *encourage l'harmonisation des politiques et pratiques réglementaires dans le domaine nucléaire, en ce qui concerne notamment la sûreté des installations nucléaires, la protection de l'homme contre les radiations ionisantes et la préservation de l'environnement, la gestion des déchets radioactifs, ainsi que la responsabilité civile et les assurances en matière nucléaire ;*
- *examine régulièrement les aspects économiques et techniques de la croissance de l'énergie nucléaire et du cycle du combustible nucléaire, et évalue la demande et les capacités disponibles pour les différentes phases du cycle du combustible nucléaire, ainsi que le rôle que l'énergie nucléaire jouera dans l'avenir pour satisfaire la demande énergétique totale ;*
- *développe les échanges d'informations scientifiques et techniques concernant l'énergie nucléaire, notamment par l'intermédiaire de services communs ;*
- *met sur pied des programmes internationaux de recherche et développement, ainsi que des activités organisées et gérées en commun par les pays de l'OCDE.*

Pour ces activités, ainsi que pour d'autres travaux connexes, l'AEN collabore étroitement avec l'Agence Internationale de l'Énergie Atomique de Vienne, avec laquelle elle a conclu un Accord de coopération, ainsi qu'avec d'autres organisations internationales opérant dans le domaine nucléaire.

The development of safe methods for disposal of highly radioactive wastes from nuclear power generation is of common interest in many countries, and is an important item in the programme of the OECD Nuclear Energy Agency. Emplacement of the wastes deep underground is a disposal option which offers the prospect of their isolation for very long periods of time. To investigate both the feasibility and long-term safety of this option, it is necessary to undertake experimental investigations in suitable rock at a depth comparable to that of a future repository. Such in-situ experiments are performed at the Stripa mine, a disused iron mine in central Sweden, and a few other locations in the OECD area.

With the support of the NEA Co-ordinating Group on Geological Disposal and the Division KBS of the Swedish Nuclear Fuel Supply Company, a workshop was held in Stockholm in October 1982 to review the results of this research in recent years. The Division KBS manages the International Stripa Project, an autonomous NEA project in which eight countries participate. The Co-ordinating Group on Geological Disposal has a mandate from the NEA Radioactive Waste Management Committee to organise the exchange of information and to co-ordinate research in this field.

These proceedings reproduce the papers contributed to the meeting. The summary and conclusions has been prepared by the NEA Secretariat in collaboration with key contributors to the workshop. Any opinion expressed is the responsibility of the authors and in no way commits Member governments of the Organisation.

Le développement de méthodes sûres d'évacuation de déchets hautement radioactifs, engendrés par la production d'électricité d'origine nucléaire, est d'un intérêt commun pour plusieurs pays et il représente un élément important du programme de l'Agence de l'OCDE pour l'Energie Nucléaire. L'évacuation par enfouissement des déchets à grande profondeur offre une perspective d'isolement pour des périodes très longues. Afin d'étudier la faisabilité et la sûreté à long terme de cette méthode, il est nécessaire d'entreprendre des investigations expérimentales dans des roches appropriées à une profondeur comparable à celle d'un dépôt souterrain éventuel. De telles expériences in situ sont actuellement réalisées à la mine de fer désaffectée de Stripa, située dans le centre de la Suède, ainsi que dans certaines autres installations des pays de l'OCDE.

Avec le soutien du Groupe de coordination de l'AEN sur l'évacuation des déchets radioactifs dans les formations géologiques et de la Division KBS de la Compagnie suédoise d'approvisionnement en combustibles nucléaires, une réunion de travail s'est tenue à Stockholm, en octobre 1982, dans le but de faire le point sur les recherches réalisées ces dernières années. La gestion du Projet international de la mine de Stripa est confiée à la Division KBS. Il s'agit d'un projet autonome de l'AEN auquel participent huit pays. Le Groupe de coordination de l'AEN a pour mandat d'organiser un échange d'informations et de coordonner les activités de recherche dans ce domaine.

Ce compte rendu comporte les communications présentées à cette réunion. Le résumé ainsi que les conclusions ont été établis par le Secrétariat de l'AEN et les principaux collaborateurs à la réunion. Les opinions exprimées n'engagent que la responsabilité des auteurs et en aucun cas les gouvernements des pays Membres de l'Organisation.

TABLE OF CONTENTS

TABLE DES MATIERES

Session 3

GEOCHEMICAL AND MIGRATION INVESTIGATIONS

RECHERCHES DANS LE DOMAINE GÉOCHIMIQUE ET DE LA MIGRATION

Chairman - Président : O. HEINONEN (Finland)

Session 4

INVESTIGATION OF BUFFER AND BACKFILL MATERIALS

RECHERCHES DANS LE DOMAINE DES MATÉRIAUX TAMPONS ET DE REMBLAYAGE

Chairman - Président : P. GNIRK (United States)

Session 5

PROGRAMMES FOR FURTHER IN SITU EXPERIMENTAL WORK

PROGRAMMES FUTURS DES TRAVAUX EXPERIMENTAUX IN-SITU

Chairman - Président : L.-B. NILSSON (KBS)

SUMMARY AND CONCLUSIONS

Granite rock formations have attracted considerable attention in many NEA Member countries because of properties that may make them potentially suitable for the isolation of highly radioactive wastes from nuclear power generation. Granite and similar crystalline rocks are relatively common in several NEA countries, and are generally very old, dating typically from many hundred of millions of years in the past. The massive and compact nature of these rocks often makes them very impermeable to water. Movement through fractures with groundwater is seen as the only, or at least the dominant, mechanism by which radioactive material could possibly return to man's environment. These hard rock formations also have great structural strength and resistance to erosion or other disruptive events. Hence, radioactive wastes placed deep in these rocks is very unlikely to be disturbed by climatic or geological events, or by accidental or intentional human intrusion.

The physical and chemical properties of hard crystalline rocks are nevertheless complex, particularly as regards groundwater movement, fracture networks, and chemistry. The presence of highly radioactive and heat-generating material in mined cavities at great depth can also modify the local environment and change the characteristics of the surrounding rock and hydrology. In order to investigate the behaviour of hard rocks in circumstances similar to those that would exist in a radioactive waste repository, it is necessary to perform in-situ experiments and measurements at similar depth to that of a conceptual repository. There are few existing mines that give access at such depths to appropriate comparable geologies, however one such opportunity is provided by the Stripa mine in Sweden.

During 1977-1980 a series of investigations in the field of radioactive waste storage was conducted in the Stripa mine. The main part of the investigations was performed by the Lawrence Berkeley Laboratory (LBL), University of California, sponsored by the US Department of Energy (DOE) in cooperation with the Swedish Nuclear Fuel Supply Company (SKBF) through the Division KBS. The aim of these experiments was to develop techniques to determine near-field rock mechanics, and far field hydrological, geochemical and geophysical parameters at potential waste repository sites.

The International Stripa Project started in May 1980 and the first phase is scheduled to continue for four years until April 1984. This phase is supported by seven NEA Member countries, five as full members (Finland, Japan, Sweden, Switzerland, USA) and two as associated members (Canada and France). Research in the Stripa mine in Sweden concerns geochemical and hydrogeological studies, migration experiments, and studies of the function of engineered barriers for geological disposal of radioactive waste in granite formations. The management of the project is entrusted to the Division KBS of the Swedish Nuclear Fuel Suply Company.

The Stripa mine is an abandoned iron mine in central Sweden. A granite formation is adjacent to the ore excavations and is accessible at a depth of 350 metres. Horizontal tunnels have been excavated into this granite where rock conditions are suitable for experimental investigations. The rock is a grey to light red, medium-grained granite. It contains several fracture sets and a normal groundwater situation. The majority of fractures are however closed and filled with chlorite and occasionally calcite.

Hydrogeological and geochemical investigations in boreholes are designed to test methods and instruments for studies in horizontal and deep vertical boreholes. Additional information is obtained about the hydraulic characteristics of granite, on interactions between fractures, and on chemical conditions in groundwater at great depth. During 1981 a heavily fractured zone was intersected by a deep vertical borehole. The hydraulic conductivity of the fractured zone, about 860 metres below the surface, is several orders of magnitude greater than that of the main part of the granite formation.

In connection with developing an understanding of radionuclide migration in fissured media, various inactive tracers representing all the important radionuclides are followed through a single fissure in which there is naturally flowing water. The experiments show how well laboratory sorption data can predict radionuclide migration, and provide information on dispersion in a single fracture. A preparatory test has been completed in 1981, and the main experiment started in 1982.

Although laboratory tests can provide information on the chemical and physical properties of buffer and backfilling materials, it is necessary to test the behaviour of an integrated system of heaters, buffer materials, host rock and groundwater in a full-scale representative environment. In the "Buffer Mass Test", electric heaters, simulating nuclear waste, are placed in holes drilled into the floor of a horizontal tunnel. The cylindrical heaters are surrounded by a buffer of highly compacted bentonite. The space above the heaters is then filled with a mixture of bentonite and sand. The heaters, the deposition holes and the surrounding rock are instrumented to monitor temperatures, moisture content, swelling pressure in the bentonite, and water pressure and displacements in the rock. The first heaters were switched on in late 1981, and the evolution of the system will be monitored until late 1984.

Related experimental work has also been done at other sites in NEA countries:

- The Climax mine in a granite formation on the US Nevada Test Site is used for rock mechanics studies. Eleven canisters of commercial light-water reactor fuel and additional electrical heater simulators are placed in boreholes beneath tunnels to study temperature distributions, displacements and stresses in granite, and to investigate any effects associated with the intense radiation from spent fuel;

- The Colorado School of Mines (CSM) under sponsorship of the US Office of Nuclear Waste Isolation has established a hard rock test facility in a mine at Idaho Springs, Colorado. This is in jointed gneiss and lies under approximately 100 meters of rock cover, but above the groundwater table;

- In the UK, a programme of field studies on heat transfer and fracture hydrology is conducted at the AERE site in Cornish granite. A heater test has been operated for 3 years and fracture hydrology is studied by pumping water into fractures from cored boreholes at 200 meters depth;

- In France, boreholes have been drilled to a depth of 500 and 1000 meters in a granite formation. These have been used for fracture assessment, hydrology tests, and geochemical monitoring.

ROCK MECHANICAL INVESTIGATIONS

The effect of heat from radioactive decay in high-level waste is an important consideration in the design and safety assessment of underground repositories. Tests in the Stripa mine, at the Climax mine in the USA and in Cornish granite in the UK have investigated the temperature and stress distributions near a heater simulating high-level waste. Three experiments to simulate the thermal effect of radioactive waste have been done at a depth of 340 meters in the Stripa mine.

One experiment involved an array of 8 electrical heaters, scaled to simulate in one year the thermal conduction field around radioactive waste canisters over a period of a decade. The other two experiments each involved a single electrical heater to simulate the short term effects in the rock. The temperature distributions have proved to be consistent with predictions on the basis of thermal conduction through the rock matrix. The observed temperatures are very close to those calculated by a linear heat conduction model. There is some evidence of slightly greater heat flow above the midplane of heaters, but this is interpreted as due to induced convection in the heater. The accurate prediction of temperatures above 100°C indicates that the phase change from water to steam in these media is not an important contributor to heat transfer.

A separate consideration is the extent to which heat would be removed by the ventilation airstream in an open mined repository. Ventilation and dewpoint measurements in the Climax mine indicate that only about 13% of the heater energy is removed. This is less than predicted, indicating that the complexities of heat transfer to the airstream are not yet adequately modelled.

Displacements associated with thermal expansion of rock have been monitored in the Stripa tests and in the Climax mine. The displacements are small, of the order of one or two millimeters over a distance of several meters near the heater. Although good agreement between measured and calculated displacements is found in the Climax tests, the measurements at Stripa have shown discrepancies between observed values and those calculated on the assumption of an unfractured monolithic rock matrix. With high rock temperatures, differences are seen between displacements adjacent to the upper and lower sections of a heater, and there is hysterisis in the thermally induced expansion and contraction cycles. These effects are attributed to irreversible movement on fractures in the rock. The closing of fractures may provide an explanation for the anomalously low values of rock displacements during heater experiments at Stripa. Thermally induced rock movements in fractured granite are small, non-linear, and not yet predictable. The phenomenology of jointed rock response to heating is not yet fully understood.

The stresses and stress changes in deep granite environments are important in excavation of cavities and in assessments of their stability. The in-situ stress at Stripa has been measured by two techniques: overcoring and hydraulic fracturing in vertical boreholes to depths of 400 meters. The vertical stress is approximately equal to the weight of the overburden, but one of the horizontal stresses is several times larger. Underground stress measurements reflect the effects of old mine excavations and the experimental drifts. These measurements have demonstrated that existing techniques can be used to give data of sufficient accuracy for repository design. Stress measurements should be made at an early stage in site characterisation. When underground workings are available higher resolution measurements should be used to follow local stress variations. The two techniques are proven and complementary:
-Hydraulic fracturing measurements show less scatter and provide an integrated value over a larger scale; overcoring gives a more

detailed and accurate stress measurement. There is however a need for further development of instruments and techniques for stress change measurements.

Investigations have been made by the Colorado School of Mines of the extent and degree of disturbance around an underground excavation. A variety of techniques such as crosshole ultrasonics and permeability and deformation measurements have been used, and emphasis has been placed on developing appropriate instrumentation. Changes in stress and damage are difficult to resolve but there is a preliminary indication of a disturbed zone extending about one meter from the excavation, with an altered permeability.

HYDROLOGICAL INVESTIGATIONS

A comprehensive programme of fracture investigations has been carried out in the granite at Stripa. Data has been taken from a limited number of outcrops, surface and subsurface boreholes, and fracture maps of rooms and drifts at a depth of about 340m. Analysis of fracture statistics from these sources has revealed the existence of four fracture sets and has illustrated the statistical distributions of fracture characteristics. A probabilistic safety analysis of a repository would require that pertinent properties of the host rock be described in terms of statistical distributions, and the experience at Stripa has indicated how fractures should be characterised and what constitutes an adequate statistical sample.

The fracture hydrology programme at Stripa was also designed to explore new techniques for evaluating the fluid flow properties of discontinuous rock. The field work at Stripa included permeability measurements over various scales, from tests on individual fractures to the very large scale macropermeability experiment. A comparison of borehole results with those of the macropermeability experiment has shown good agreement with an average hydraulic conductivity of the order of 10^{-10} m/s.

The hydrogeology investigations at Stripa have demonstrated the value of co-ordinated surface and underground testing in site characterisation. These both provide complementary information, and show that the full characterisation cannot be obtained from one approach alone. The aim of the hydrogeological investigations in boreholes has been to acquire a better understanding of deep-lying bedrock and groundwater. The programme has included geophysical logging, hydraulic tests, rock stress measurements, and hydrogeochemical sampling. Experience gained in the design and testing of instruments has been the most valuable aspect of these investigations. Hydrogeological models show, as expected, a considerable influence of the mine on the flow pattern even for the deepest boreholes, hence some hydrogeological information obtained so far is very site specific and of limited generic application. The drilling programme has however demonstrated that core-drilling can be carried out with only slight deviations from predicted orientations. The vertical boreholes have shown that fracture zones with hydraulic conductivity of greater than 10^{-8} m/s exist even at great depth. Further work on the development of underground investigation methods should focus on non-destructive cross-hole techniques to characterise large volumes of rock. One such hydrological testing method using sinusoidal pressure variations has been developed by the UK Institute of Geological Sciences. In the UK it has been shown that a signal can be detected over a distance of 45 meters in fractured granite. This approach shows some advantages over conventional testing methods and will be further developed in tests at Stripa.

HYDROGEOCHEMICAL ANALYSES

Hydrogeochemical analyses of groundwater from the deep boreholes at Stripa show an increase in salinity and pH with depth accompanied by a decrease in alkalinity and sulphate concentration. It is hoped that interpretation of these results will lead to generic insights into the origin and distribution of salinity in deep groundwaters of crystalline rock masses. Groundwaters have been investigated with geochemical and isotopic techniques with particular attention given to discharge from the fracture system at 800 meters depth. The elevated salinity of these waters may be related to admixture with fossil sea water or with leaching of old fluid inclusions from the genesis of the granite. Isotopic analyses show conclusively that the waters are old and originate in an environment which differs markedly from the modern environment at Stripa. The deep waters however have a complex history, and it has so far proved impossible to quantify all sources and processes which participated in their genesis. Despite this some general remarks are valid: it is very likely that a significant amount of mixing occurs between different waters, thus the determination of absolute water ages is not appropriate; regional processes appear to dominate over granite reactions in the geochemical evolution of the waters (the mixing of fresh and marine waters and geochemical reactions with minerals and fracture fillings); the combination of isotopic and geochemical analyses can discriminate between flow systems and can contribute to the hydrogeological characterisation of a site.

MIGRATION IN FRACTURES

A pilot experiment and preliminary work for investigation of migration in a single fissure of tracers simulating radionuclides has been completed at a depth of 360 meters at Stripa. Tracers are injected under pressure, and water samples are collected under anoxic conditions at a distance of several meters from the injection point. The pilot experiment has shown that most water flows in a few separated channels in a fissure rather than uniformly over the fissure area. The travel times in separate channels can be different even for non-sorbed tracers, and the flow rates of the channels can differ considerably. For the main investigation, with a migration distance of about five meters, the connecting fissure has been located by the hydraulic pulse technique. It is not always where expected from geometric projection of the visible fissure in the drift and from core and TV observations in drill holes. It is clear from this work that groundwater pathways are considerably more complex than implied by a single fissure concept. The pressure, flow and tracer observations indicate that there is a large difference between visible fissure widths and those expected from pressure drop predictions on the basis of parallel-wall flow. The fissure widths predicted from pressure drop data we considerably smaller.

Modelling of tracer migration in other experimental sites by the UK has shown the importance of diffusion into stagnant pore water, fracture fillings or the rock matrix as a retention mechanism even for non-sorbed tracers.

THE BUFFER MASS TEST

The large scale buffer mass test in Stripa is concerned with the validation of predictive modelling of the temperature, pressure and water uptake into compacted bentonite around electrical heaters and into a sand/bentonite mixture used to refill the tunnel above the heater assemblies. Installation of all components of the test was completed successfully in early 1982, and all heater assemblies have been operating for over six months. The temperature fields in the deposition holes reached an almost steady state after approximately one month. The recorded temperatures are well within the predicted range. The water uptake in the heater holes, as interpreted from moisture gauge signals, has taken place in reasonable agreement with predictions. This confirms that the pattern of water bearing fractures in the rock largely determines the rate and distribution of the water uptake. The moistening of the tunnel backfill is also so far in fairly good agreement with predictions. It has been demonstrated that the backfill has remained in good contact with the rock roof of the tunnel. Water saturation of the bentonite and backfill is taking place but calculations indicate that full saturation of the entire backfill will not be achieved during the test. The bentonite swelling pressure in the deposition holes has shown a steady increase, correlated with water uptake. The most interesting phase of the test started in late 1982 with excavation of the compacted bentonite from the deposition holes. A careful sampling of water content will allow a more accurate validation of model predictions.

FUTURE EXPERIMENTAL INVESTIGATIONS

A second phase of experimental work in the Stripa project started in 1983. This involves development of cross-hole geophysical investigation techniques; sealing of boreholes, shafts and tunnels; a large scale 3-dimensional tracer migration test; and fracture hydrology tests. Underground laboratories are also under development in granite areas in Switzerland and Canada, a close collaboration between these programmes and the Stripa Project is ensured. Brief descriptions of these programmes are given in the final session of this workshop.

FIGURE 1

WHAT HAS BEEN LEARNED FROM INVESTIGATIONS AT STRIPA?

- Heat transfer in fractured rocks is by conduction and is predictable
- Thermally induced rock movements are non-linear and are not yet predictable
- There is a need for further instrument development for stress change measurements
- Accurate fracture mapping is required in order to model:
  - Thermo-mechanical rock responses
  - Groundwater movement in the flow system
- Geochemical and isotopic analyses of groundwaters provide corroborative evidence of site impermeability
- Migration of tracers by groundwater flow follows complex pathways in channels within fractures
- Preliminary results indicate that the behaviour of bentonite as a buffer material is in agreement with predictions
- Large-scale underground tests are a necessary complement to laboratory experiments.

FIGURE 2

ADDITIONAL INFORMATION THAT COULD BE OBTAINED FROM THE INVESTIGATIONS AT STRIPA

- A complete description of the geological setting and history of the Stripa granite formation;
- A demonstration of the applicability of inclined boreholes in site selection;
- A complete investigation of the response of rock to temperature and pressure changes;
- A complete analysis of thermomechanical response data;
- A complete analysis of the macropermeability test;
- Development of new methods for evaluating rock behaviour, including effects of rock stress and fracture development;
- A complete analysis of the Buffer Mass Test, the migration experiments, and hydrogeological investigations.

Session 1

ROCK MECHANICAL INVESTIGATIONS

Chairman - Président

O. STEPHANSSON

(Sweden)

Séance 1

RECHERCHES EN MATIERE DE LA MECANIQUE DES ROCHES

PROGRESS WITH THERMOMECHANICAL INVESTIGATIONS OF THE STRIPA SITE

N.G.W. Cook,+ P.A. Witherspoon,+* E.L. Wilson+ and L.R. Myer*
+University of California, Berkeley
*Lawrence Berkeley Laboratory
Berkeley, California
USA

ABSTRACT

The Stripa mine in Sweden has provided an unusual opportunity to investigate the thermomechanical behavior of a granitic rock mass and the effect on such behavior of the discontinuities that are a persistent feature. Experiments to simulate the thermal effects of burying radioactive wastes have been carried out at a depth of about 340 m. All of the thousands of underground measurements of temperatures, displacements, and stresses in the rock, recorded continuously over this period, have been made available on digital tape and in hard copy. No complete analysis nor interpretation of these data has yet been made; this paper presents some results of recent investigations.

Although temperature fields predicted by calculations based on linear heat conduction compare well with measured field data, evidence for some convective heat transfer has been found. Measured values of displacements have been significantly less than those predicted by calculations based on linear thermo-elasticity for intact rock. Evidence that these displacements are diminished by fractures in the granite has been found. A program of laboratory investigations has been started to measure coefficients of thermal expansion and elastic moduli for Stripa granite. Intact rock samples recovered from boreholes in the vicinity of the heater experiments have been tested over a range of temperatures from 20°C to 200°C and hydrostatic stresses from 2 to 55 MPa. The results provide a sound basis for predicting thermomechanical behavior, but it will also be necessary to carry out similar measurements on fractured samples in order to adequately understand how discontinuities affect rock behavior.

The final goal of these thermomechanical investigations is to develop a capability for predicting the overall structural behavior of an underground repository. A project has been started to develop an Underground Structural Analysis Program - USAP, which is a special version of the SAP-80 program. USAP is not only an effective program for the correlation studies associated with the Stripa project, but has the potential to be used effectively in far more general problems associated with the underground storage of nuclear waste. The options of temperature-dependent nonlinear materials, consideration of incremental excavation and construction sequences, multilevel substructures, large capacity and efficient numerical methods make USAP a unique finite element program that has been developed specifically for the analysis of underground structures.

INTRODUCTION

The Stripa mine in Sweden has provided an unusual opportunity to investigate the thermomechanical behavior of a granitic rock mass and the effect on such behavior of the discontinuities that are a persistent feature. Another paper in this workshop describes the results of investigations on the hydrogeology at the Stripa site [1] and includes the geological setting and the work that was performed in characterizing the fracture system. The paper that follows presents a review of progress that has been made with thermomechanical investigations.

Three experiments to simulate the thermal effects of burying radioactive wastes have been done at a depth of about 340 m in a granite (quartz monzonite) mass in the Stripa mine [2]. The objectives and design of these experiments were described at an earlier meeting of the OECD Nuclear Energy Agency [3]. One experiment involved an array of 8 electrical heaters, scaled so as to simulate in one year the heat conduction field around radioactive waste canisters over a period of a decade. The other two experiments each involved a single, full-scale electrical heater to simulate the short-term, proximate effects in the rock of heat, such as would be released by the radioactive decay of nuclear wastes. One of the full-scale experiments used a power output of 3.6 kW, corresponding to a canister of high-level nuclear wastes 5 years out of the reactor and the other an output of 5 kW, corresponding to a canister of nuclear wastes 3 years out of the reactor. Each heater measured 0.3 m in diameter by 3 m in length and was buried in a vertical hole 0.406 m in diameter by 5.5 m deep in the floor of an underground drift.

ANALYSIS OF THERMOMECHANICAL MEASUREMENTS

Data Collection

A comprehensive suite of instruments was installed in order to measure thermomechanical effects in each of the three heater experiments. Four types of instruments were used: (1) thermocouples for temperature measurement, (2) U.S. Bureau of Mines (USBM) borehole deformation gauges for stress determination, (3) IRAD vibrating-wire gauges for stress determination, and (4) rod extensometers for displacements. Details of the installation and calibration of these instruments have been reported [4].

Changes in temperatures in the rock around the two, full-scale heaters were measured with thermocouples installed at elevations from 3.0 m above to 3.0 m below the mid-plane of each heater in six vertical boreholes at radial distances from 0.4 m to 0.9 m from the axis of the heater. Relative displacements between points in the rock adjacent to each heater were measured using 15, 4-point wire extensometers located in nine horizontal and six vertical boreholes. The USBM deformation gauges and IRAD "stress meters" were also installed in the same rock to assess changes in stress. The heaters were operated at full power for 400 days, during which measurements were recorded continuously, followed by another 200 days of recording after the heaters had been switched off.

Predicted changes in temperatures, displacements and stresses over the heating period were calculated, using the theory of linear thermoelasticity, in advance of the experiments [5]. During the period of heating, comparisons were made between measured and predicted values. In general, linear heat conduction was found to yield a good prediction of the measured temperature distributions. However, the measured displacements between extensometer anchor positions consistently proved to be about half the displacements predicted from the theory of linear thermoelasticity; predictions were based on a value of $\alpha = -11.1 \times 10^{-6} °C^{-1}$ for the linear coefficient of thermal expansion for the rock, which proved to be too high.

The suite of measurements made underground at Stripa has been documented fully [6] but a complete analysis of these data has not been made yet. This paper presents a partial analysis of data for those situations in which it is comparatively easy to do the necessary calculations by hand.

Temperatures

At a radial distance of about 0.7 m from the full-scale heater axes, temperatures in the mid-plane of the 3.6 kW heater were below 100°C and those of the 5 kW heater were above 100°C. Comparisons between measured temperatures and calculated, conductive temperatures at these positions enable one to assess the effects, if any, of phase changes from water to steam on the temperature field. Similarly, comparisons between measured temperatures and calculated, conductive temperatures for positions at this radial distance and 1.5 m above and below the mid-plane of the heater enable one to assess the effects of convection on the temperature field.

In Figure 1 is shown a comparison between calculated, conductive temperatures and measured temperatures in the vertical thermocouple hole T15, at a radial distance of 0.69 m from the axis of the 3.6 kW heater, for the heating and cooling cycle. Position C is in the mid-plane, B is about 1.5 m above this plane and D is about 1.5 m below it. Linear regression analyses of the data show that the correlations between the measured and calculated temperatures are high, $r > 0.998$. This result was obtained using a linear relationship between calculated temperature, x, and measured temperature, y, in the form:

$$y = mx + b, \qquad (1)$$

where m is the straight line slope and b is the intercept on the axis for measured temperature. From the parameters given on Figure 1, it is evident that the intercepts are relatively small, but the slopes are quite variable. As the theory of heat conduction is linear, the effects of non-linearities, such as thermal convection, should result in finite intercepts on the axes, or correlation coefficients less than unity. Slopes other than unity, with correlation coefficients near unity and relatively small values for the intercepts, indicate that the values of the theoretical constants are not correct, although linear theory is applicable. The lowest of the three lines in Figure 1, T15D, has the largest value for the intercept, but this value is still relatively small. All these lines have high correlation coefficients. This suggests that heat flow in the rock outside the heater hole was by conduction, and the flow above the mid-plane was greater than that below it. Investigations by others [7] confirm that induced convection in the heater caused more heat to flow above the mid-plane than below it.

Comparisons between calculated values of temperatures at a radial distance of 0.7 m from the axis of the 5 kW heater for three similar vertical locations relative to the mid-plane are shown in Figure 2 for thermocouple hole T21. All of the correlation coefficients are significantly less than unity, and the values of the axial intercepts for positions T21B and T21D are relatively high. This suggests that some heat flows other than by conduction. However, if anything, the data are closer to being linear for temperatures above 100°C than they are for temperatures below 100°C, indicating that phase changes from water to steam are not an important part of these non-linearities.

Displacements

Comparisons between measured and calculated values of the relative displace-ements between points about 2.2 m above and below the mid-planes of the heaters and at radial distances of about 2 m from the heater axes are shown in Figure 3. Extensometer hole E7 is adjacent to the 3.6 kW heater and E12 is adjacent to the 5 kW heater. Measured relative displacements at E7 are plotted as a function of calculated displacements to investigate the effects of heating and cooling at the 3.6 kW heater. For the 5 kW heater, only displacements at E12 during heating for the first 200 days are plotted; at that time eight peripheral heaters uniformly separated around a circle at a radial distance of 0.9 m from the 5 kW heater were switched on to simulate the ambient heating that would occur as a result of the interaction of canisters in an array. The calculation of the effects of this additional heating followed by cooling cannot readily be calculated by hand.

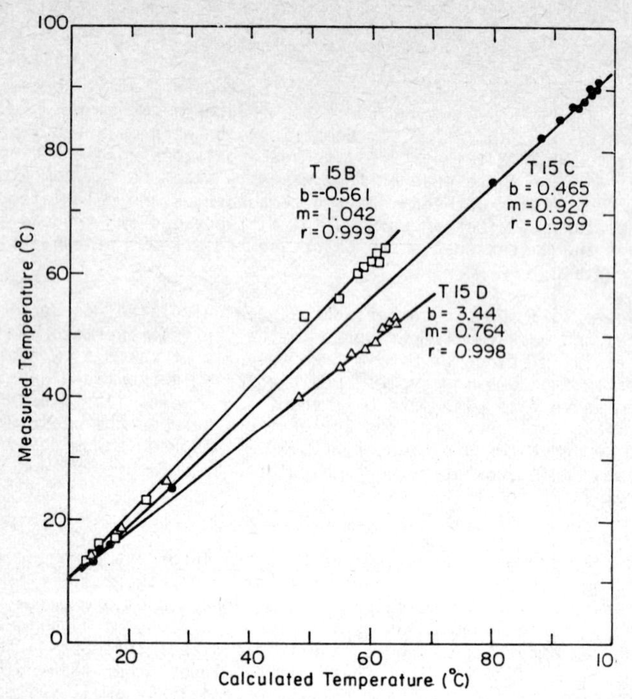

Fig. 1. The relationships between calculated and measured temperatures in the rock around the 3.6 kW heater at a radius of 0.69 m and three different vertical locations: 1.5 m above (B), on (C), and 1.5 m below (D) the mid-plane of the heater.

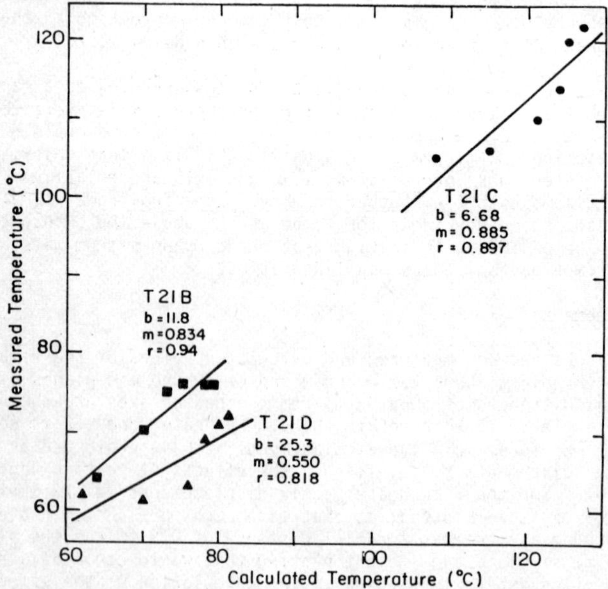

Fig. 2. The relationships between calculated and measured temperatures in the rock around the 5 kW heater at a radius of 0.7 m and three different vertical locations: 1.5 m above (B), on (C) and 1.5 m below (D) the mid-plane of the heater.

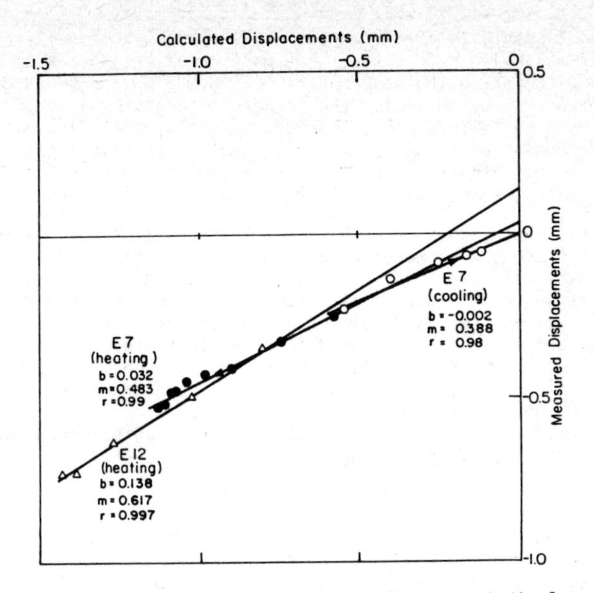

Fig. 3. The relationships between calculated and measured displacements between positions in the rock about 2.2 m above and below the mid-planes of the heaters, as a function of the mean temperature of the rock between these positions.

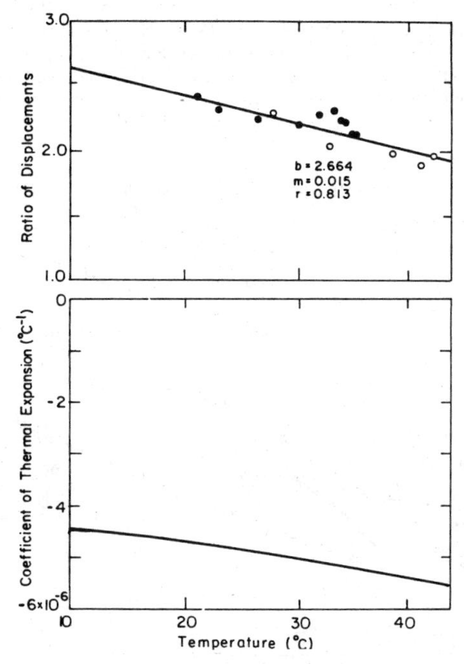

Fig. 4. The ratio between calculated and measured relative displacements of positions in the rock about 2 m from the heater axes and about 2.2 m above and below the mid-planes of the heaters as a function of the mean temperature of the rock between these positions, and inferred in situ values for the linear coefficient of thermal expansion.

The correlation coefficients for the two sets of heating data and one set of cooling data are all close to unity, $r > 0.997$. However, the intercepts increase systematically with increasing temperature, b = 0.032 mm for the 3.6 kW heater during heating and b = 0.138 mm for the higher temperatures produced during heating with the 5 kW heater. This suggests that the coefficient of thermal expansion is dependent on temperature. Also, note that all slopes are much less than unity which indicates that the numerical value of the coefficient used in our predictive calculations ($-11.1 \times 10^{-6}{}^{\circ}C^{-1}$) was much too large.

Temperatures in the rock between these extensometer anchors were also measured and predicted, and therefore a value can be derived for the in situ coefficient of thermal expansion. The ratio between the calculated and measured displacements is shown in Figure 4 as a function of the measured mean rock temperature ($\overline{T}$) between the extensometer anchors. The mean rock temperature adjacent to the 3.6 kW heater increased for the first 400 days and then decreased for the remaining 200 days. The mean temperature adjacent to the 5 kW heater increased by an amount (5 kW/3.6 kW) greater than that around the 3.6 kW heater for the first 200 days, until the peripheral heaters were switched on. The ratio, R, between the calculated and measured displacements can be approximated (correlation coefficient $r = 0.813$) by a line:

$$R = 2.66 - 0.015\,\overline{T} \tag{2}$$

over a range of mean temperatures $20°C < \overline{T} < 45°C$. The value of the in situ linear coefficient of thermal expansion α^*, can be found by dividing the value used in the calculations by R, that is,

$$\alpha^* = \alpha/R \tag{3}$$

Values of the in situ coefficient α^* as a function of mean temperature, $\overline{T}$, are shown in Figure 4. Note that the rock mechanics convention in use here makes expansion and the coefficient of thermal expansion both negative.

Hysteresis of the thermally induced expansion and contraction cycles can conveniently be studied using data obtained in the E7 extensometer adjacent to the 3.6 kW heater. One of the extensometer anchors, B, was located close to the mid-plane of the heater. The calculated and measured relative displacements between this anchor and those about 2.2 m above and below the mid-plane, A and C respectively, are as is illustrated in Figure 5, for the heating and

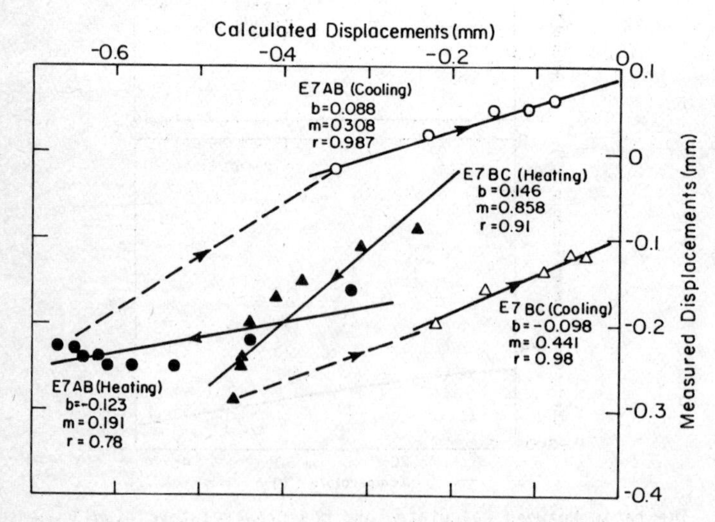

Fig. 5. The relationships between calculated and measured relative displacements of positions in the rock at a radial distance of about 2 m from the axis of the 3.6 kW heater and at three different vertical locations: 2.2 m above (A), on (B), and 2.2 m below the mid-plane of the heater during heating and cooling periods.

the cooling periods of the experiment. The overlapping plots are somewhat confusing, but the purpose is to illustrate how the heating and cooling effects for the upper section, AB, differ from those of the lower section, BC. Note first of all that the correlation coefficients for both the heating and cooling periods for lower interval, BC, are near unity but that their slopes are quite different from one another. For the upper interval, AB, the correlation coefficient was near unity only for the cooling period.

However, over the heating interval the fitted lines between calculated and measured displacements have significant values for the intercepts, b, indicating that the thermal expansion is decidedly non-linear. For the cooling period, both intervals also have finite intercepts, but this may have resulted from the non-linearity during heating. Also over the cooling period, the slopes for both the AB and BC sections are similar, but the values range from 0.308 to 0.441 indicating again that the value for the coefficient of thermal expansion used in our predictive calculations was much too large.

The non-linearity in measured displacements could have been caused by small discontinuities (cracks) in the rock mass [2]. Thermally induced expansions as measured between anchors positioned on either side of a crack would appear to be diminished because the expanding rock would first move to close the opening and not necessarily produce a measureable displacement. Conversely displacements between positions that do not include a crack would appear to be accentuated as a result of a diminished restraint from the adjacent rock because of the crack. Such is the situation in Figure 5. The displacements between positions B and C over the heating period are accentuated, that is, they have the greater slope; whereas those measured between positions A and B are diminished, that is, they have the smaller slope. This seems to suggest that a crack (or cracks) in the rock mass between positions A and B affected the displacements significantly.

During the cooling period, the data for section BC tend to support the same concept of a discontinuity within the rock mass between A and B. The BC intercept of -0.098 mm indicates the effect of a permanent set having occurred as a result of the fracture first closing and then not recovering the same total displacement upon opening. This is commonly observed in laboratory testing of fractured rocks [8]. Thus, it appears that the closing of fractures may provide an explanation for the anomalously low values of the thermally induced rock displacements during the heater experiments at Stripa.

Stresses

The effect of fractures in the rock, such as those needed to explain the small values of the thermally induced displacements, would result in lower values of the thermally induced stresses than would be predicted by calculation. Stiff inclusion "stressmeters" and borehole deformation gauges were installed in the heater experiments. The data are stored as frequencies of the vibrating wire for the IRAD "stressmeters" and as changes in borehole diameter for the USBM deformation gauges. However, the amount of calculation involved precludes the analysis of these data by hand.

The virgin state of stress at Stripa has been measured by overcoring and hydraulic fracturing in a vertical borehole from surface to a depth of 381 m [9]. It appears that the vertical component of the virgin stress is approximately equal to the weight of the overburden and that one of the horizontal stresses is about twice the vertical component. There is some evidence to suggest that the state of stress observed in this borehole is perturbed from the virgin state of stress in proximity to the mine. Other stress measurements have been made underground in the vicinity of the heater experiments [10, 11]. The underground stress measurements reflect the effects of stress concentrations around the old mine excavations and around the experimental drifts. The proper analysis and interpretation of these data require extensive numerical modeling of the stress distributions around the openings.

MATERIAL PROPERTIES INVESTIGATIONS

The above discussion reveals the need for accurate data on material properties in any investigation on the thermomechanical behavior of a fractured rock mass. The comparisons of measured displacements with calculated displacements that were based on constant values for the thermoelastic parameters clearly reveal the need for a better definition of these parameters both in terms of absolute values and in terms of any non-linearities. Therefore a program of laboratory investigations has been started to measure coefficients of thermal expansion and elastic moduli for Stripa granite as a function of pressure and temperature. The program is not yet complete but the progress that has been made to date can be summarized as follows.

The rock samples chosen for laboratory testing were selected from cores recovered from instrumentation holes close to those in which measurements of displacements and stresses were made. The samples are thus from the same rock mass that has been subjected to the heater experiments. All samples tested to date have been intact. It will be necessary to investigate fractured samples in order to better understand how discontinuities affect the material properties, but this has not yet been done.

To bracket temperature and stress conditions that existed during the heater experiments, tests were performed over a range of temperatures (T) from 20°C to 200°C and hydrostatic stresses (confining pressure, P) from 2 to 55 MPa. Maximum deviatorial stress varied depending on conditions from 60 MPa at T = 200°C and P = 2 MPa, to 260 MPa at T = 20°C and P = 55 MPa.

To minimize the number of samples that had to be tested, each sample was subjected to a matrix of pressure-temperature states and the test sequence was designed to minimize sample damage [12]. Property measurements started using the highest confining pressure, and a heating and cooling cycle was completed at each level before proceeding to the next lower pressure. The properties measured at each pressure and temperature were: (a) tangent Young's modulus, E_T; (b) Poisson's ratio, ν; (c) volumetric coefficient of thermal expansion, α_V; and the (d) linear coefficient of thermal expansion, α_ℓ. The details of this work have been reported by Myer and Rachiele [12].

Major trends in the thermal expansion behavior of intact Stripa granite are illustrated in Figure 6. This figure presents summary curves for average values of the linear coefficient of thermal expansion, α_ℓ. Note the isobaric increase in the values of α_ℓ with increasing temperature; values at 20°C were 52% to 70% of those at 180°C. The curves on this figure are nearly linear, indicating that the rate of increase in α_ℓ can be considered constant over the range of temperatures experienced in the heater experiments. Pressure effects, on the other hand, were much less. At any given temperature, increasing confining pressures from 2 to 55 MPa resulted in slight decreases in α_ℓ, from 5% to 10%. Essentially the same results were obtained for α_V, and the details can be found elsewhere [12]. These results are consistent with those of Cooper and Simmons [13] and Heard and Page [14].

The laboratory results for the linear coefficient of thermal expansion have confirmed the non linearity in α_ℓ as deduced from the analysis of field data. At temperatures below 100°C and at hydrostatic stresses of about 15 MPa (i.e. the approximate conditions in the rock mass immediately around the heaters), α_ℓ is less than -11.1×10^{-6}°C^{-1}, the value used in the predictive calculations for thermally induced displacements. However, the values shown on Figure 6 are still higher than those computed from an analysis of field data. As shown on Figure 4, the field derived value of α_ℓ is of order -5×10^{-6}°C^{-1}, whereas the laboratory result at comparable conditions of P and T is of order -8×10^{-6}°C^{-1}. This discrepancy between in situ and laboratory results may well be explained by the influence of crack closure on rock displacments.

Effects of changes in hydrostatic stress on the tangent Young's modulus, E_T, are illustrated in Figure 7. Because the moduli also varied as a function

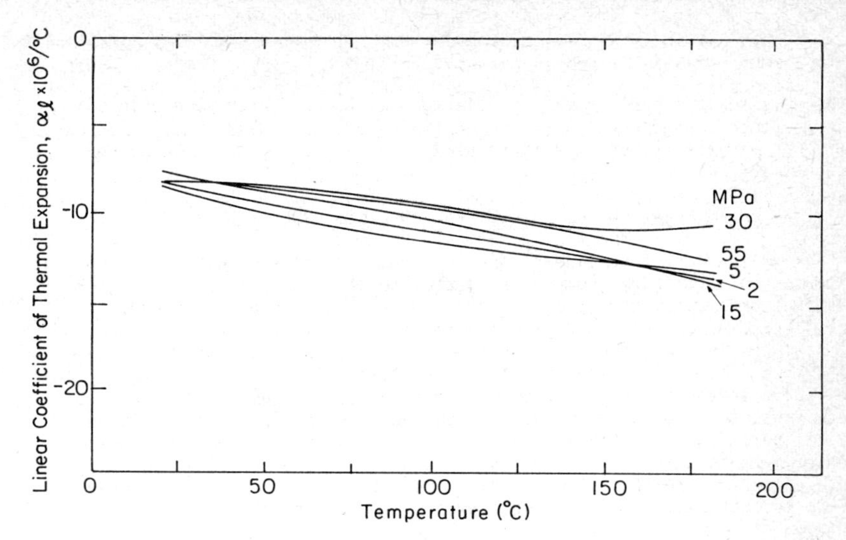

Fig. 6. Average values of linear coefficient of thermal expansion as a function of temperature for several confining pressures.

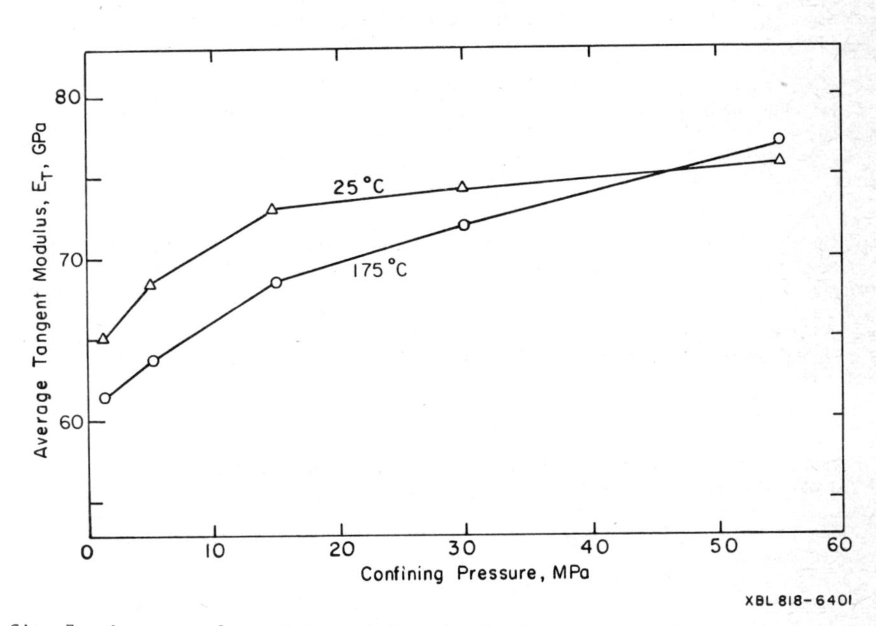

Fig. 7. Average values of tangent Young's Modulus as a function of confining pressure at 25°C and 175°C. All values obtained with deviatorial stress of 60 MPa.

of deviatorial stress, all data have been referred to a common deviatorial stress of 60 MPa. Note the isothermal increases in E_T; values at 2 MPa were between 75% and 87% of those at 55 MPa. The major part of this hydrostatic stress effect occurred when confining pressures were below 15 MPa.

Results for Poisson's ratio, ν, averaged about 0.22 with some evidence of temperature dependence. When the confining pressure was 2 MPa, ν decreased about 26% over the temperature range used; whereas, at 55 MPa, the decrease was about 13%.

DEVELOPMENT OF SPECIAL PURPOSE STRUCTURAL ANALYSIS PROGRAM-USAP

The evaluation of nonlinear displacements and stresses within underground rock masses of arbitrary geometry subjected to thermomechanical loading can be estimated by a computer program based on the finite element method. Several existing computer programs for nonlinear analysis were evaluated for potential use in the correlation studies associated with the Stripa heater tests. Because of the unique nonlinear material properties and the requirements to evaluate stresses due to incremental excavation, it was not possible to directly use these programs without extensive modifications. In addition, most of the existing nonlinear analysis programs were based on numerical and computer methods which were over ten years old and were on the verge of obsolescence. Therefore, it was decided to extend a new computer program, SAP-80, to satisfy the special requirements of the Stripa project.

The SAP-80 Program

The recently developed SAP-80 computer program [15] is based on modern computational techniques and contains the modularity and flexibility which allows modifications to be rapidly incorporated and verified. Following is a partial list of some of its most important options:

1. All program segments interact with a common file data base. Therefore, pre and post processors are an integral part of the program. Also, restart capability is built into the program.

2. Extensive mesh generation and three-dimensional mesh plotting are possible.

3. The data input files are of an unique free format which are designed to be prepared on interaction computer terminals or by the traditional card format.

4. The equation solver is of the blocked profile type and the program contains an automatic profile minimization routine.

5. Multilevel substructure (or super element) options are incorporated in the program and are easily activated.

6. Isoparametric plane or axisymmetric elements with from three to nine nodes have been included in the program.

7. The program has been designed to operate effectively on modern super minicomputers.

Development of USAP

The develpment of the Underground Structural Analysis Program - USAP is a special version of the SAP-80 program. To date, approximately one man-year has been devoted to USAP. A two-dimensional version of the program has been developed and tested. The following options have now been verified:

1. Realistic temperature dependent nonlinear material properties have been incorporated in the program and the solution at each point in time is obtained by iteration.

2. Each element has a "birth" and "death" time. Therefore, stresses at different stages of excavation or construction can be obtained.

3. The substructure (or super element) option allows areas of the mesh representing the surrounding rock mass that are far removed from the high stress regions to be modeled by elastic substructures in which all the unknowns are eliminated within these substructures. As a result, only a limited number of elements are required in the iterative nonlinear analysis (See Figure 8).

4. Geometric constraints have been added to the program which allow coarse and fine meshes to be merged without the use of triangular elements. Therefore, a consistently higher level of accuracy is obtained (see Figure 8A).

Prior to the termination of this study, work had begun on the development of a three-dimensional element (4 to 27 nodes) with temperature dependent, nonlinear properties. This effort is about half completed, and work was also initiated on the development of a general three-dimensional mesh generation program for underground structures with openings of arbitrary geometric shape.

It has been demonstrated that USAP is not only an effective program for the correlation studies associated with the Stripa project, but has the potential to be used effectively in far more general problems associated with the underground storage of nuclear waste. The options of temperature-dependent nonlinear

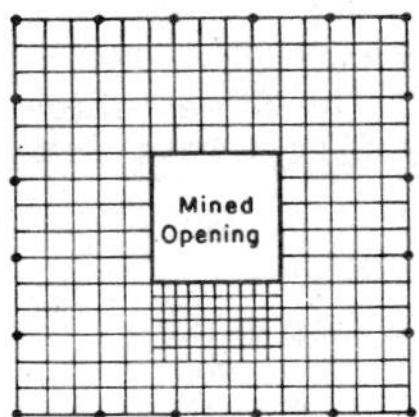

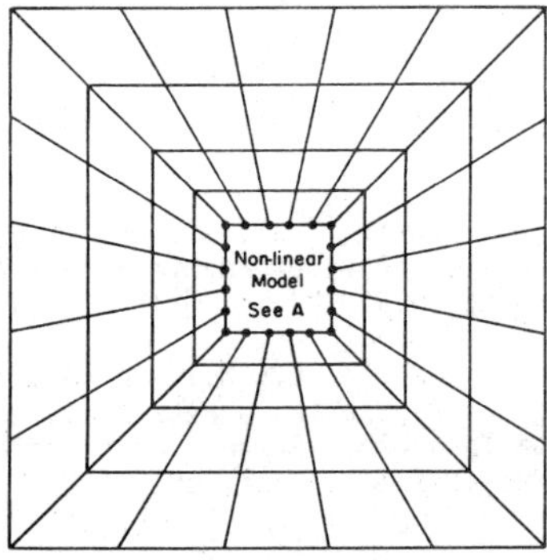

Fig. 8. USAP mesh showing nonlinear finite element model (A) for the mined openings surrounded by a condensed elastic substructure model (B) representing the surrounding rock mass in the far field. Note arbitrary location of mesh refinement within non-linear model.

materials, consideration of incremental excavation and construction sequences, multilevel substructures, large capacity and efficient numerical methods makes USAP a unique finite element program that has been specifically developed for the analysis of underground structures. However, significant work is still required to document the use of the program and to add full three-dimensional analysis capability.

ACKNOWLEDGMENT

This work was supported by the Assistant Secretary for Nuclear Energy, Office of Waste Isolation of the U.S. Department of Energy under contract DE-AC03-76SF00098. Funding for this project is administered by the Office of Nuclear Waste Isolation at Battelle Memorial Institute.

REFERENCES

1. Gale, J.E., Witherspoon, P.A., Wilson, C.R., and Rouleau, A., "Hydrogeological Characterization of the Stripa Site," paper presented at OECD Nuclear Energy Agency Workshop on "In-Situ Experiments in Granite Associated with Geological Disposal of Radioactive Waste," Stockholm, Sweden, Oct. 25-27, 1982.

2. Witherspoon, P.A., Cook, N.G.W. and Gale, J.E., "Geological Storage of Radioactive Waste: Field Studies in Sweden," Science, 211, pp. 894-900, 1981.

3. Cook, N.G.W. and Witherspoon, P.A., "In Situ Heating Experiments in Hard Rock: Their Objectives and Design," paper presented at OECD Nuclear Energy Agency Workshop on "In Situ Heating Experiments in Geologic Formations," Ludvika, Sweden, Sept. 12-15. 1978.

4. Schrauf, T., Pratt, H., Simonson, E., Hustrulid, W., Nelson, P., DuBois, A., Binnall, E. and Haught, R., "Instrumentation Evaluation, Calibration, and Installation for the Heater Experiments at Stripa," Lawrence Berkeley Lab. Rep. LBL-8313, SAC-25, 1979.

5. Chan, T., Cook, N.G.W. and Tsang, C., "Theoretical Temperature Fields for the Stripa Heater Project," Lawrence Berkeley Lab. Rep. LBL-7082, SAC-09, 1978.

6. Chan, T., Binnall, E., Nelson, P., Stolzman, R., Wan, O., Weaver, C., Ang, K., Braley, J. and McEvoy, M., "Thermal and Thermomechanical Data from In Situ Heater Experiments at Stripa, Sweden," Lawrence Berkeley Lab. Rep. LBL-11477, SAC-29, 1980.

7. Javandel, I. and Witherspoon, P.A., "Thermal Analysis of Stripa Heater Test Data - Full Scale Experiment," Lawrence Berkeley Lab. Rept. LBL-13217, (in preparation) 1982.

8. Witherspoon, P.A., Wang, J.S.Y., Iwai, K., and Gale, J.E., "Validity of Cubic Law for Fluid Flow in a Deformable Rock Fracture," Water Resources Res., 16, 6, pp. 1016-1024, 1980.

9. Doe, T., Ingevald, K., Strindell, L., Haimson, B., and Carlsson, H., "Hydraulic Fracturing and Overcoring Stress Measurements in a Deep Borehole at the Stripa Test Mine, Sweden," Proceedings "22nd U.S. Symposium on Rock Mechanics," Massachusetts Institute of Technology, pp. 373-378. 1981.

10. Carlsson, H., "Stress Measurements in the Stripa Granite," Lawrence Berkeley Lab. Rept. LBL-7078, SAC-04, 1978.

11. Doe, T.W., Hustrulid, W., Leijon, B., Strindell, L., Ingevald, K. and Carlsson, H., "Results and Conclusions of the Stripa In Situ Stress Measurement Program," paper presented at OECD Nuclear Energy Agency Workshop on "In-Situ Experiments in Granite Associated with Geological Disposal of Radioactive Waste," Stockholm, Sweden, Oct. 25-27, 1982.

12. Myer, L.R. and Rachiele, R., "Laboratory Investigations of Thermomechanical Properties of Stripa Granite, Part I: Apparatus, Part II: Application and Results," Lawrence Berkeley Lab. Rep. LBL-13435, (in press) 1982.

13. Cooper, H.W. and Simmons, G., "The Effect of Cracks on the Thermal Expansion of Rocks," Earth and Planetary Science Letters, 36, pp. 404-412, 1977.

14. Heard, H.C. and Page, L., "Elastic Moduli, Thermal Expansion and Inferred Permeability of Two Granites to 350°C and 55 MPa," J. Geophys. Res., (in press) 1982.

15. Wilson, E.L., "SAP-80 Structural Analysis Program for Small or Large Computer Systems," Proceedings CEPA 1980 Fall Conference, Newport Beach, Calif., October 13-15, 1980.

RESULTS AND CONCLUSIONS OF STRESS MEASUREMENTS AT STRIPA

T. W. Doe
Lawrence Berkeley Laboratory
University of California
Berkeley, California, U.S.A

W.A. Hustrulid
Colorado School of Mines
Golden, Colorado, U.S.A

B. Leijon
University of Luleå
Luleå, Sweden

K. Ingevald, L. Strindell
Swedish State Power Board
Vällingby, Sweden

Hans Carlsson
Swedish Nuclear Fuel Supply Co. (KBS)
Stockholm, Sweden

ABSTRACT

This paper describes the results of stress measurements at Stripa, compares the results obtained by different techniques, and recommends a stress measurement program for a hard rock repository site. The state of stress at the Stripa Mine has been measured both in a 381 m deep hole drilled from the surface and in holes drilled from the drifts underground. Hydraulic fracturing and several overcoring methods have been used (Lulea triaxial gauge, CSIRO gauge, USBM gauge, Swedish State Power Board deep-hole Leeman triaxial gauge). The results of overcoring and hydraulic fracturing agree well, particularly for the magnitude and orientation of the greatest stress. A recommended program for stress measurement at a repository site would include hydraulic fracturing and deep-hole overcoring in a deep hole drilled from surface, and overcoring (Lulea gauge and USBM gauge) and hydraulic fracturing from holes drilled from underground openings when access is available. Propagation of the hydraulic fractures should be monitored acoustically to determine their location and orientation.

1. INTRODUCTION

Over the past several years stress measurements have been performed at the Stripa Mine in conjunction with the rock mechanics and hydrologic tests performed at the site. As a result of this work, there is now an extensive data base which is useful both for analyzing the results of the experiments that have been performed at the site and for developing stress measurement programs for other hard rock repository sites.

This paper serves three purposes. First, it summarizes the results of all the stress measurements that have been performed at the site. Second, it presents a comparison of the results by different methods and in different locations around the mine. Third, it makes suggestions as to how stress measurement programs can be designed for other hard rock repository sites. Due to limitations of space, detailed descriptions of the techniques used and the results are not included in this paper. This information is presented by Doe and others [1].

2. HISTORY OF STRESS MEASUREMENT ACTIVITIES

The first stress measurements were performed by Carlsson [2] in 1977 as part of the University of Luleå heater tests. The Leeman triaxial gauge was used for the measurements which were performed in a 20m hole drilled from the Luleå drift (Figure 1).

In 1980, an extensive stress measurement program was undertaken as part of the LBL-KBS Swedish-American Cooperative (SAC) project. The first phase of this program was to determine the stresses at a distance where the mine effects would be negligible. A 381 m vertical borehole, SBH-4, was drilled about 400 m north of the experimental area (Figure 1). The Swedish State Power Board performed stress measurements using their unique deep-hole Leeman gauge at hundred meter intervals over the length of the hole. After the drilling was complete, stress measurements were carried out by hydraulic fracturing. This effort represented the first time hydraulic fracturing and overcoring had been carried out in a common deep hole.

The second phase of the SAC program work was a series of stress measurements performed in the area between the Extensometer and Full Scale Heater test drifts underground (Figure 1 and 2). A vertical hole, BSP-1, was drilled in the floor of the Full Scale drift for measurements using hydraulic fracturing and the Swedish State Power Board Leeman gauge. Horizontal holes BSP-2 and BSP-3 were drilled from the Extensometer drift under the Full Scale drift for hydraulic fracturing and overcoring. BSP-1 and 2 were 76 mm in diameter; BSP-3 was 150 mm in diameter. The overcoring methods included the CSIRO triaxial gauge, the University of Luleå (LuH) triaxial gauge, and the USBM borehole deformation gauge.

The most recent stress measurements have been made in borehole V1 which is collared at the 360 meter level of the mine [3]. The Power Board has performed two sets of four measurements each at hole depths of 150 and 300 meters. These measurements are the deepest that have been made at the site.

3. FAR FIELD MEASUREMENTS

3.1 Stress Measurement Data

Measurement of the stresses in deep hole SBH-4 has been described elsewhere [4]. The Power Board triaxial gauge has been adapted from the Leeman triaxial gauge for use in deep holes by wireline emplacement. The gauge measures the complete state of stress through the response to overcoring of three strain gauge rosettes, each having three components. The rosettes are cemented to the wall of a 36 mm pilot hole which is then overcored with a conventional 76mm (NX) double tube core barrel. The data exhibit a large degree of scatter in the magnitudes [4]; however, there is consistency in the orientations of the principal stresses. The greatest principal stress is oriented horizontally. but,the other principal stresses are generally skewed with respect to the vertical and horizontal. Hence the usual assumption in hydraulic fracture data analysis that the borehole is oriented in the direction of one of the principal stresses is not met.

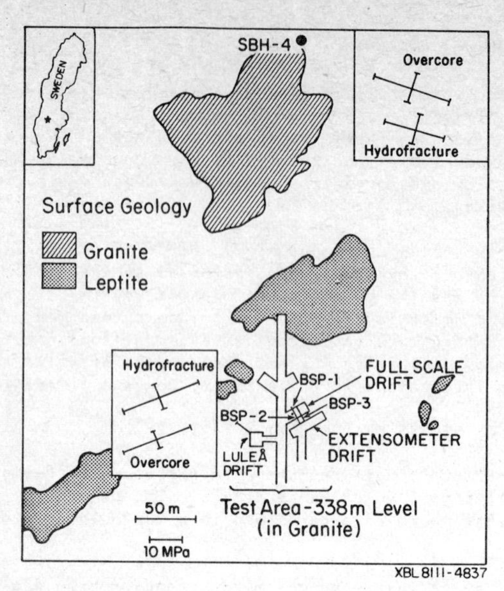

Figure 1. Map of Stripa Mine showing location of test areas and stress measurement holes.

The methods used to interpret the hydraulic fracturing records are described in detail in Doe and others [4]. Briefly. the methods use the first breakdown pressure and a tensile strength term determined in laboratory testing. This method is considered more reliable than second breakdown techniques [5,6] for sites where the ratio of the horizontal stresses exceeds two [4], as for such ratios the theoretical second breakdown pressure is less than the shut in pressure. The tensile strength term has been derived using methods of statistical fracture mechanics [7] which take into account the differences of size effect and sample geometry between laboratory tests and field fracturing tests. The orientations of the fractures were obtained using a wireline impression packer which contained a borehole survey compass for packer orientation.

3.2 Comparison of the Far-Field Hydraulic Fracturing and Overcoring Results

The results of the overcoring and hydraulic fracturing have been compared based on the orientation of the maximum horizontal stress, and the magnitudes of

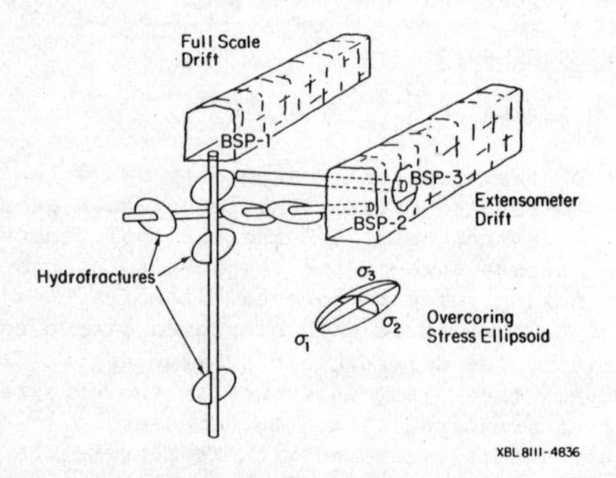

Figure 2. Cartoon of Full Scale drift area showing locations of stress measurement holes, orientations of hydraulic fractures, and approximate overcoring stress ellipsoid.

the maximum and minimum horizontal stresses at a depth of 320 m in the hole. This is the approximate depth of the test facility. The horizontal stresses are used for comparison because the hydraulic fracture test is generally thought to measure mainly the stress components normal to the borehole. The true stress magnitude at the test facility depth is estimated by interpolation of a linear regression of stress versus depth.

The orientation of the maximum horizontal stress depth is shown as a function of depth in Figure 3. The mean values of 9 hydraulic fractures and 11 overcores below 200 m stress direction agree within a one degree of N 83 W. The 95% confidence levels for the means are determined using the methods of Mardia [8] and are both about ± 20 degrees. Thus one can conclude that the correspondence between the overcoring and hydraulic fracturing is quite good. The confidence intervals could have been improved to about ± 15 degrees had over twenty measurements been made. Further improvement in the statistics with larger numbers of measurements would probably not be practical from the standpoint of cost and from the lack of suitable test zones.

The magnitudes of the secondary principal stresses for the overcoring and the hydraulic fracturing are shown as a function of depth in Figures 4 and 5. The data have been fitted to regression lines whose coefficients are given in the figures. The horizontal stress magnitude for the two methods interpolated to the depth of the test facility agree closely (Table 1). The hydraulic fracturing has somewhat better confidence intervals than the overcoring, particularly for the horizontal minimum stress, but both methods provide estimates for the mean stress values at the depth of the test facility within ± 20% or better.

The stress data by both methods is highly variable as shown by the values for standard errors of estimate for the regression and the confidence interval values for the slopes of the regression lines (Fig. 4 and 5). Despite these large values, the confidence intervals for the interpolations are srelatively small because a large number of measurements were made. One can conclude from these data that reliable predictions of the in situ stresses at depth cannot be made either on the basis of a few measurements or by extrapolating the results of a set of measurements taken at shallow depth.

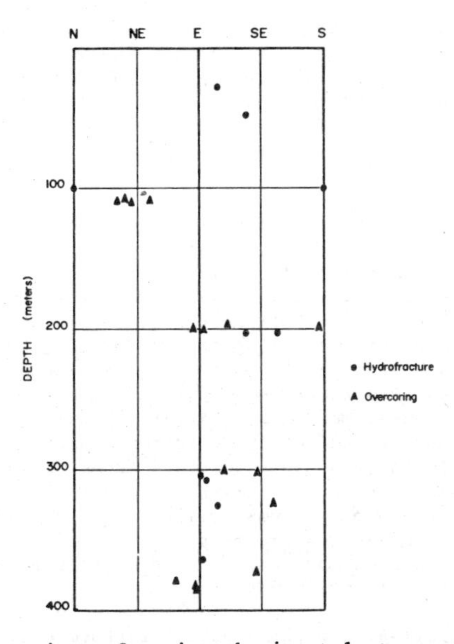

Figure 3. Measured orientations of maximum horizontal stress in SBH-4.

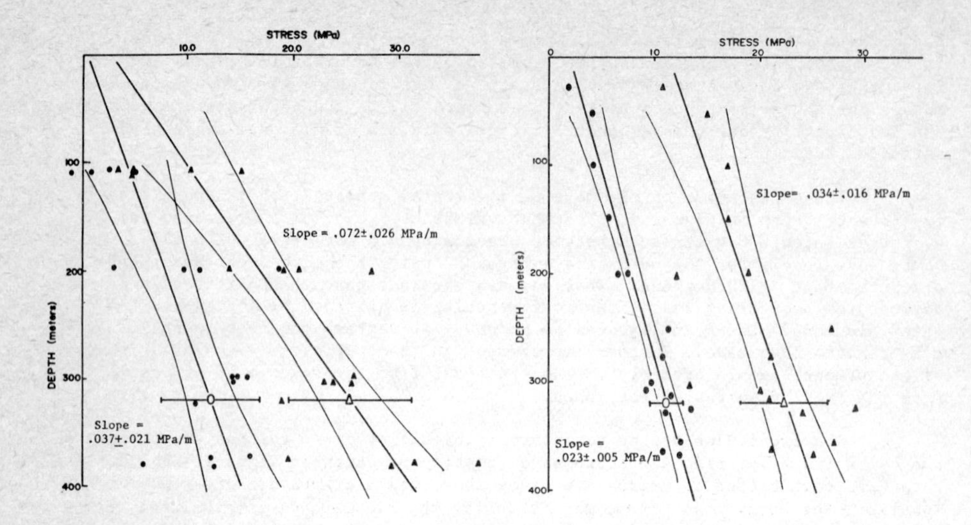

Figure 4. (left) Horizontal secondary stresses in SBH-4 as determined by overcoring. Curved lines are 90% confidence limits for regression line ordinate; bar at test facility depth (320 m) is standard error of estimate for regression.

Figure 5. (right) Horizontal secondary stresses in SBH-4 as determined by hydraulic fracturing; see Figure 4 for explanation.

4. NEAR-FIELD MEASUREMENTS

4.1 Measurements in the Luleå Drift

Carlsson [2] performed a series of 19 stress measurements in a borehole drilled off of the Luleå drift (Figure 1). The measurements were performed using the Leeman triaxial gauge. The overcored hole had an 87 mm diameter and the pilot holes were 38 mm in diameter.

The orientations of the individual points and the mean orientations of the principal stresses are shown in Figure 6. The mean values and orientations of the principal and secondary stresses are given in Table I. Contrary to the SBH-4 results, the maximum stress in the Luleå drift is oriented northeast with a plunge to the northwest.

4.2 Measurements in the Full Scale Drift Area

The second phase of the LBL stress measurement program was to measure the in situ stress in the immediate vicinity of the the full scale heater experiment (Figures 1 and 2). Three holes were drilled for the purposes of stress measurement. BSP-1 was drilled vertically downward from the center line of the full scale drift to a depth of 25 m. This hole was 76mm in diameter and was used for hydraulic fracturing and for overcoring by the Power Board method. Two holes were drilled from the extensometer drift, an opening excavated parallel to the full scale drift at a lower level to allow the installation of horizontal extensometers in the original heater experiment. Hole BSP-2 was drilled with a diameter of 76mm to a length of 20 m and was used exclusively for hydraulic fracturing. The hole was drilled at an angle three degrees downward from the horizontal to assure that the hole would remain full of water during the hydraulic fracturing tests. Hole BSP-3 had a diameter of 150 mm and was drilled to a length of 12 m for use in USBM, CSIRO, and LuH triaxial gauge measurements. It was drilled at a small angle upward from the horizontal to assure that water would drain from the hole and not affect the bonding of the triaxial strain gauges.

An acoustic emission experiment was performed by Ernest Majer of Lawrence Berkeley Laboratory to detect the propagation of the hydraulic fractures and,

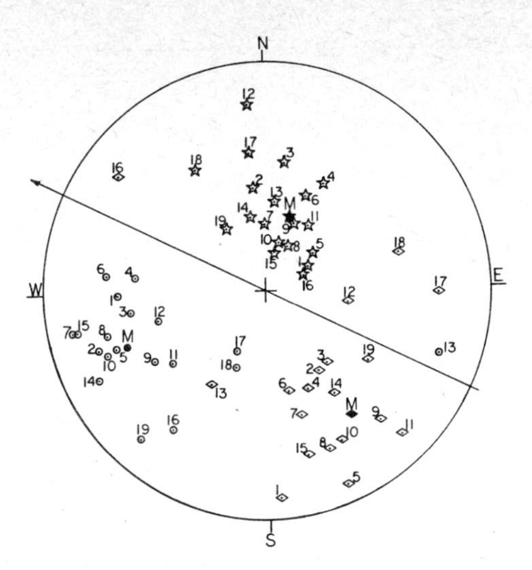

Figure 6. Principal stresses measured by overcoring off Luleå drift. Circle-maximum stress, diamond- intermediate stress, star-minimum stress. Mean values denoted by "M". Lower hemisphere stereographic projection.

hopefully, to map their locations. The layout and results of the acoustic experiment are discussed in Majer and McEvilly (1982).

In addition to the simple comparison of stress values from the various overcoring techniques, the underground experiment had several other objectives including:

o investigating the effect of hole orientation on the hydraulic fracture results,

o measuring the influence of the extensometer drift and full scale drifts on the in situ stress orientations and magnitudes,

o investigating the correspondence of the acoustically mapped hydraulic fracture plane with the plane normal to the least principal stress determined by the overcoring.

4.3 Power Board Leeman Gauge Measurements (BSP-1)

A total of six measurements were made with the Power Board Leeman gauge in BSP-1. The measurements were made between 1.3 and 10.0 m below the floor of the full scale drift. The mean principal stress data are given in Table I and the orientation data are shown in Figure 7. The orientation of the maximum stress is very consistent among the measurements and is oriented northeast-southwest, parallel to the axes of the two drifts. The intermediate principal stresses are oriented off the vertical an average of about 30 degrees to the southeast. The minimum principal stresses are within about 30 degrees of the horizontal.

4.4 LuH Triaxial Gauge Measurements (BSP-3)

The University of Luleå triaxial gauge is an adaptation of the Leeman gauge used in the Luleå drift measurements. The major differences in equipment and procedures relate to the cleaning of the hole to assure good bonding of the strain gauges and use of four component strain gauge rosettes. Eight LuH triaxial gauge measurements were made at depths between 2.5 and 11.2 m in BSP-3. The magnitudes of the principal stresses are given in Table I and the orientations are shown in Figure 8. The maximum principal stress is consistently parallel to the axes of the drifts and coincides closely with the direction measured by the Power Board. The intermediate and minor principal stresses are nearly 45 degrees off the vertical

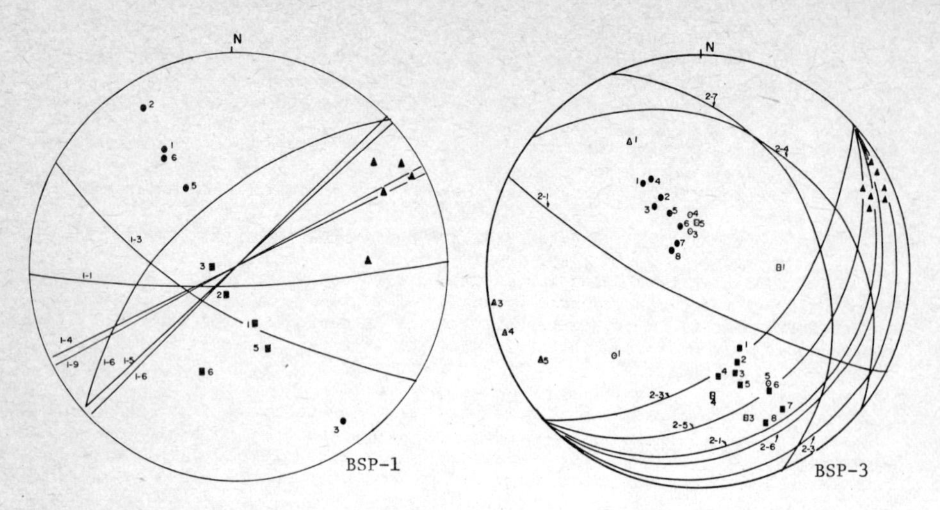

Figure 7. Principal stresses measured by Power Board overcoring and orientation of hydraulic fractures in BSP-1. Triangle- maximum stress, square- intermediate stress, circle- minimum stress. Lower hemisphere stereographic projection.

Figure 8. Principal stress measured by LuH gauges (solid symbols) and CSIRO gauges (open symbols) in BSP-3 with orientation of hydraulic fracture planes in BSP-2. See Figure 7 for explanation of symbols.

and horizontal directions near the collar of the hole. As the hole proceeds toward the full scale drift, the intermediate stress rotates toward the horizontal and the least stress rotates toward the vertical. Near the drifts the minimum stresses will be normal to the drift walls, hence one would expect the minimum stress to rotate away from horizontal toward vertical along the length of the hole.

4.5 USBM Borehole Deformation Gauge Measurements (BSP-3)

The USBM borehole deformation gauge was used in the same hole as the LuH gauge and CSIRO gauge measurements. Unlike the triaxial strain gauges, the USBM gauge measures only the stress components normal to the hole axis. This disadvantage is balanced against the greater rapidity and reliability of the USBM gauge. Triaxial gauge and deformation gauge measurements complement one another when used in the same hole. The triaxial gauges provide the three dimensional information, and the deformation gauge provides the larger number of measurements necessary for confidence in the stress determination for a site.

Nine USBM measurements were made at hole depths ranging from 1.1 to 9.7 m. The results of the USBM measurements are plotted along with the secondary stress data for the LuH gauge measurements in Figure 9. The mean stress values are given in Table I. The agreement for both magnitude and orientation is excellent. The orientation of the maximum secondary stress is horizontal for both techniques.

4.6 CSIRO Triaxial Gauge Measurements (BSP-3)

The CSIRO triaxial gauge [9] is a hollow cylinder which is grouted into a 38 mm pilot hole and then overcored. The gauge is similar to the Leeman triaxial gauge in that it contains three strain gauge rosettes with three components each. The data reduction methods are the same as those for the Leeman gauge except for modifications to allow for the effect of the cylinder. The CSIRO gauge has several practical advantages over the Leeman gauge including protection of the electronic circuitry from the drilling fluids and capability for monitoring the strain gauge outputs during the overcoring. It has a disadvantage in that the cements require seventeen hours or more to cure to an acceptable hardness and the gauge is not as reliable in water filled holes.

Five CSIRO measurements were made in BSP-3. Despite using curing times in excess of seventeen hours, the first two measurements indicated inadequate bonding to the pilot borehole walls. Even after switching to a faster curing cement for the final three measurements, the gauge values showed an average drift rate of about five microstrains per minute before and after the overcoring. The mean orientation and magnitude data are calculated using strain data from which the linear drift has been subtracted and are presented in Table I. The data, shown in Figure 8, are consistent with the LuH results both in orientation and in magnitude.

4.7 Near Field Hydraulic Fracturing Measurements (BSP-1 and BSP-2)

Hydraulic fracturing stress measurement were carried out in the vertical borehole BSP-1 and the horizontal borehole BSP-2. Nine measurements were carried out in BSP-1 over 0.6 m test intervals ranging in depth from 2.3 m to 20.2 m. Eight measurements were performed in BSP-2 using the same test interval length at depths of between 3.8 m and 16.7 m.

The equipment and procedures used for conducting the tests and evaluating the results were essentially the same as those used for the far field stress measurement work in SBH-4. The results, given in Table I, are calculated using the first breakdown pressures and the tensile strength values determined by Ratigan [6]. It was assumed that the underground test area was drained of water, thus the pore pressure term was taken as zero.

The orientation of the hydraulic fractures was determined by an impression packer which was lowered into the hole on scribed tubing. Figure 7 shows the orientation of the hydraulic fracture planes at the borehole wall for the vertical hole, BSP-1, and Figure 8 shows the fracture orientations for the horizontal hole, BSP-2. Hydraulic fractures propagate from the borehole in the direction of the maximum secondary stress. The fracture orientations in BSP-1 are strongly aligned parallel to the axis of the full scale and extensometer drifts. Thus the maximum stress direction determined by the hydraulic fracturing in BSP-1 agrees closely with the maximum principal stress direction determined by the overcoring measurements in both BSP-1 and and BSP-3. The hydraulic fractures in BSP-2 were horizontal rather than vertical as in BSP-1. This direction nonetheless is consistent with the BSP-1 results and the overcoring as the maximum stress is also horizontal.

The pressure-time records for the hydraulic fracturing showed distinctly different shut in pressures for the early and late pumping cycles. This difference

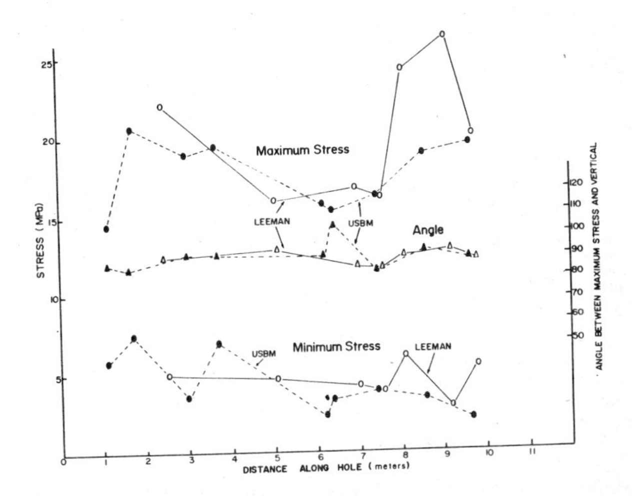

Figure 9. Secondary stresses along BSP-3 measured by LuH and USBM gauge overcoring.

was observed in both BSP-1 and BSP-2. The interpretation generally assigned to this phenomenon is that the borehole does not lie in the plane of the minimum stress, thus the fracture rotates as it propagates from the hole. The early shut in pressure would therefore represent the minimum stress normal to the hole and the later shut in pressure would be the minimum principal stress [4,5]. Following this interpretation, one would deduce that neither BSP-1 nor BSP-2 follows a principal stress direction. Were this the case one of the two boreholes would have to lie in the plane of the minimum stress and only one shut in pressure would be observed for that hole. The fact that two shut in pressures were observed in both holes, and the corresponding shut in pressures have the same value in each hole, suggests that the plane normal to the least principal stress lies 45 degrees between the two holes. This conclusion is in accord with the overcoring results, but the fact that the first shut in pressures are higher than the corresponding secondary stress values from the overcoring, and the second shut in pressures are higher than the overcoring's minimum stress values, casts some doubt on the hydraulic fracture interpretations.

The propagation of the hydraulic fractures from BSP-1 and BSP-2 was monitored with an array of acoustic sensors placed on the floor of the Full Scale drift and in the heater test instrument boreholes. Due to damage to the recording equipment in shipment, only one test from BSP-1 yielded results. Analysis of the acoustic records showed that the fracture propagated assymetrically from the borehole in a direction rotated about 20 degrees from the axis of the Full Scale drift. The most distant emissions from the fracturing were located about 2 m from the borehole. As only three stations recorded the emissions, the data was insufficient to accurately define the fracture plane, thus it is not possible to use the acoustic data to validate the stress measurement results. The fact that the emissions did appear as discrete, locatable events indicates that the method has great promise for further application.

4.8 Power Board Leeman Gauge Measurements (V1)

Measurements were performed with the Power Board's deep hole Leeman gauge at hole depths of 150 and 300 m (510 and 660 m below surface). The tests were run in sets of four. The average values of the magnitude and orientation are given in Table I. Although there is some scatter in the results, the overall magnitudes and orientations are similar to those measured in the Full Scale and Luleå drift areas. Again the maximum stress direction appears to be oriented to the northeast (Figure 10).

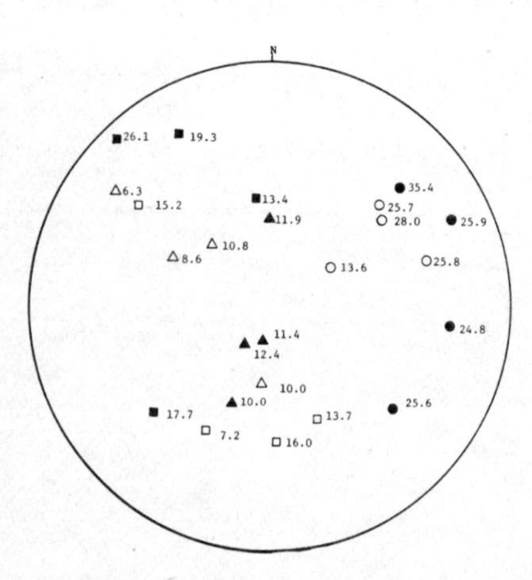

Figure 10. Orientations and magnitudes of principal stresses measured in V1 by Power Board overcoring. Circle- maximum stress, square- intermediate stress, triangle- minimum stress. Solid symbols- 150 m depth; open symbols- 300 m depth. Lower hemisphere stereographic projection.

Table 1. Average values of principal and secondary stresses measured at Stripa. Values given with 90% confidence interval.

	Principal Stresses (MPa)			Secondary Stresses* (MPa)		
	σ_1	σ_2	σ_3	σ_{Max}	σ_{Min}	σ_{Ax}
SBH-4** Hydrofrac	-	-	-	22.1±2.1	11.1±0.8	-
SBH-4** Power Board	-	-	-	25.4±2.9	12.1±2.4	-
BSP-1 Hydrofrac	-	-	-	24.0±2.9	7.6±1.0	5.1±0.8
BSP-1 Power Board	24.2±5.0	10.0±1.9	1.9±1.6	23.0±4.5	4.8±1.1	9.5±0.8
BSP-2 Hydrofrac	-	-	-	22.3±1.9	7.6±0.5	5.7±0.7
BSP-3 LuH	20.8±3.1	9.2±1.1	1.9±1.6	20.2±3.2	4.3±0.7	-
BSP-3 CSIRO	18.7±5.5	8.0±3.4	2.6±1.2	18.3±6.0	5.1±3.2	-
BSP-3 USBM	-	-	-	18.3±1.7	4.4±1.2	-
Luleå Drift Leeman Cell	19.5±2.9	8.0±3.2	4.8±1.2	15.6±2.8	8.7±1.9	10.4±1.9
V1-150m Power Board	27.9±5.0	19.1±5.3	11.4±1.0	26.0±5.1	18.3±5.9	13.4±1.4
V1-300m Power Board	22.7±6.3	13.0±4.0	9.2±2.4	19.7±7.1	11.7±3.3	13.6±1.4

* Max and Min are the stresses normal to the borehole, Ax is the stress along the borehole axis. Ax is vertical except for BSP-2, BSP-3, and Luleå Drift.

** Interpolated values at depth of test facility (338 m level).

4.9 Comparison of Near Field Results

The agreement between the results of the overcoring and the hydraulic fracturing for the Full Scale drift area measurements is excellent in the magnitude and orientation of the maximum principal stress. All the techniques are in agreement that that the direction of the maximum stress is horizontal and parallel to the axes of the full scale and extensometer drifts. The magnitudes for the stresses cover a range within about + 15% of 22 MPa.

The values for the magnitudes of the intermediate and least stresses are in general agreement; however, a number of inconsistencies exist in the orientation results. These have been discussed above, and can be summarized as (1) the inconsistency in the secondary shut in pressures between the tests run in the two orthogonal holes, and (2) the divergence in orientation between the LuH and the Power Board methods for the measurements made underneath the Full Scale drift.

The results of stress measurements from the Luleå drift and V1 also show that the maximum principal stress trends to the northeast. The overcoring results in both areas are somewhat scattered with respect to both magnitude and orientation, nonetheless the mean values are consistent with those obtained in the Full Scale drift area.

5. COMPARISON OF NEAR FIELD AND FAR FIELD RESULTS

One of the striking aspects of the comparison of the near field and far field stress data is the change in orientation of the maximum principal stress from northwest in SBH-4 to northeast in the Full Scale drift area (Figure 1). The cause of this rotation is not clear. However, as the structure of the orebody and the mine have a northeast trend, it is likely that the rotation is related to either the mine openings or the contrast in mechanical properties of the leptite and the granite. The fact that the far-field measurements are consistent with one another, as are the near-field, suggests strongly that the rotation of the stresses is real and not an instrumentation induced error. Solving the cause of the stress distribution would require stress calculations for the mine in three dimensions -- a very complicated undertaking. Chan and others [10] have prepared a two dimensional model to look the influence of the mine on the measurements at SBH-4. The model was two dimensional and completely removed the orebody as a single slab. The results showed that even with these extreme geometries the mine only influenced the stresses in the upper portions of the hole. As with most underground openings, the stress effects shown by the model die out rapidly with distance.

6. CONCLUSIONS FOR DESIGN OF STRESS MEASUREMENT PROGRAMS

The experience that has been obtained in stress measurement at the Stripa site can be used to develop recommendations for stress measurement in other hard rock locations. The site characterization program for a repository site should consider in situ stress as one criterion for acceptance of a site. Also, stress measurement data should be used to design the initial shaft and underground workings. Designers have sometimes used, in the absence of data, calculations of the the in situ stress stresses based on gravitational loading alone. This approach clearly would be in error at Stripa as the horizontal stresses are about three times the stress of the vertical load. The initial stress measurements for a repository should be made in a vertical hole similar to SBH-4. Stress should be measured at least as deep as the zone of interest, and preferably a hundred meters deeper. Measurements should be made during drilling using overcoring method like that of the Power Board. Hydraulic fracturing should be carried out after the hole is completed. Concern has been expressed that the hydraulic fractures might unduly increase the permeability of the rock, however this concern is unfounded. The acousitc results suggest that hydraulic fractures for stress measurement are limited to a few meters in size, and the permeabilities of the fracture, as shown from the shut in pressure records, are small compared to natural fractures which will be present at the site. The hydraulic fracturing complements the overcoring in that the stress measurements show less scatter and provide a larger scale, more representative value of the minimum stress magnitude (shut in pressure) and the maximum stress orientation. The overcoring complements the hydraulic fracturing in that it gives the complete state of stress and indicates if the principal stresses

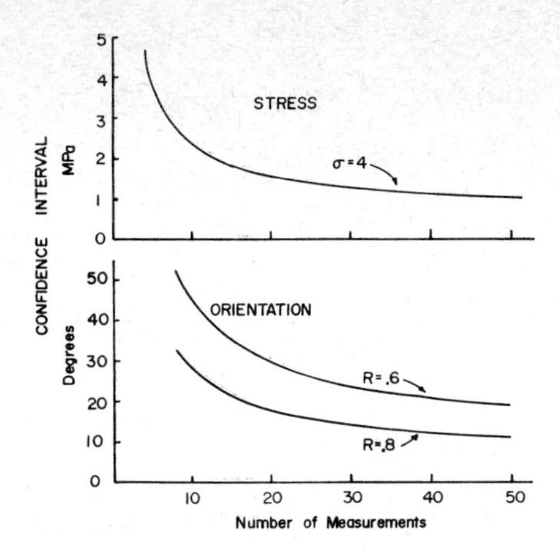

Figure 11. Improvement in confidence interval with number of measurements. Top- 90% confidence interval of stress magnitude for standard deviation of 4 MPa. Bottom- 95% confidence interval of maximum secondary stress orientation for vector length of 0.6 and 0.8.

are strongly skewed with respect to the borehole axis. Figure 11 shows the improvement in confidence interval for stress magnitude and orientation for the variance values obtained in SBH-4. The results suggest that at least 20 measurements by each technique should be made to obtain reasonably tight confidence intervals for the stress values. If a stress value at a particular depth is desired, then the measurements should either be clustered at that depth, or the measurement points should be spread over a range well above and below the depth of interest to obtain the best confidence intervals.

Once the initial underground workings have been excavated, the stress measurements should be repeated. There are two reasons for this. First, the comparison of the Full Scale drift and the SBH-4 results show that stress orientations can change significantly over distances of hundreds of meters, and second, the underground measurements have less scatter. At Stripa the most successful underground overcoring measurements were made with the USBM gauge and the LuH gauge. These can be run in the same hole and the results complement one another well. The USBM gauge is rapid to run, allowing a statistically significant sample to be taken, but it does not give the complete stress field from a single hole. The LuH gauge required more effort in bonding the gauges, but gives the complete stress field. Ten measurements by each technique should be sufficient to define the stress field within acceptable bounds. Again, the hydraulic fracturing can be used to complement the overcoring to provide a larger scale measurement. The Stripa results showed that acoustic methods have considerable potential to confirmation of the overcoring results on a large scale.

As the repository is developed additional stress measurements should be performed, particularly if anomalous structures or lithologies that might affect the stress field are encountered.

7. ACKNOWLEDGEMENTS

Mats Holmberg and Arne Tarikka assisted in the overcoring measurements in BSP-3. Mats Andersson, Karl-Ake Sjoberg, and Juri Martna participated in the Swedish State Power Board work. Per-Axel Halen's help was invaluable in setting up the details of the work at the mine. Bezalel Haimson assisted with the the SBH-4 field work and reviewed the interpretations of the tests; Milton Moebus deserves the credit for the successful performance of the hydraulic fracturing equipment in the field. Joe Ratigan provided the the laboratory tensile strength values used in the analysis of the experiment. Michael Lemcoe of Battelle Memorial Institute provided useful oversight from the Office of Nuclear Waste Isolation.

The SBH-4 and Full Scale drift work were supported by the Assistant Secretary for Nuclear Energy, Office of Waste Isolation of the U.S. Department of Energy under contract number DE-AC03-76SF0098. Funding for the project was administered by the Office of Nuclear Waste Isolation at the Battelle Memorial Institute.

8. REFERENCES

1. Doe, T.W., ed.: "Determination of the State of Stress at the Stripa Mine, Sweden", Lawrence Berkeley Laboratory Report, in preparation.

2. Carlsson, H.: "Stress Measurements in Stripa Granite", Lawrence Berkeley Laboratory Report, LBL-7080/SAC-04, 13p., 1978.

3. Strindell. L. and M. Andersson: "Measurement of Triaxial Stresses in Borehole V1", SKBF/KBS Stripa Project Internal Report 81-05, 12p., 1981.

4. Doe, T., K. Ingevald, L. Strindell. B. Haimson, and H. Carlsson: Hydraulic Fracturing and Overcoring Stress Measurements In a Deep Borehole at the Stripa Test Mine, Sweden", Proc. 22d U.S. Symposium on Rock Mechanics, Massachusetts Inst. of Tech., p. 373-378., 1981.

5. Zoback, M. and D. Pollard: "Hydraulic Fracture Propagation and the Interpretation of In Situ Stress Measurements", Proceedings 19 U.S. Rock Mechanics Symposium, University of Nevada--Reno., 1978.

6. Haimson, B.: "The Hydraulic Fracturing Stress Measurement Method and Recent Results", International Journal of Rock Mechanics, v. 15, p. 167-178., 1978.

7. Ratigan, J. :" A Statistical Fracture Mechanics Approach to the Strength of Brittle Rock", Ph.D. thesis, University of California-Berkeley, 92p., 1981.

8. Mardia, K. : "Statistics of Orientation Data", Academic Press, London, 357 p., 1972.

9. Worotnicki. G. and R. Walton: "Triaxial `Hollow-Inclusion´ Gauges for Determination of Stresses In Situ.", Proc. Int. Soc. Rock Mech. Symposium -- Investigation of Stress in Rock, Inst. of Engineers, Australia, National Conf. Pub. No. 76/4., 1976

10. Chan, T., Guvanasen, V., and Littlestone, N.: "Numerical Modelling to Assess Possible Influence of Mine Openings on Far-Field In Situ Stress Measurements", Lawrence Berkeley Laboratory Report, in press.

EXPERIMENTAL AND CALCULATIONAL RESULTS FROM THE SPENT FUEL TEST-CLIMAX*

W. C. Patrick, L. D. Ramspott, and L. B. Ballou
Lawrence Livermore National Laboratory
Livermore, California, U.S.A.

ABSTRACT

The Spent Fuel Test-Climax (SFT-C) is being conducted under the technical direction of the Lawrence Livermore National Laboratory for the U. S. Department of Energy. The SFT-C is located 420 m below surface in the Climax granitic intrusive. Following site development, 11 spent-fuel assemblies were placed in test storage in April and May 1980. At the same time, 6 electrical simulators and 20 guard heaters were also energized to begin the elevated-temperature phase of the test.

Data related to heat transfer, thermomechanical response, radiation dose, and radiation damage have been collected and are presented here, as appropriate, with calculational results. In general, measured and calculated results compare well.

RESUME

Le Spent Fuel Test-Climax (SFT-C) se fait sous la direction technique du Lawrence Livermore Lab pour le U.S. DOE. Le SFT-C se trouve 420m sous la surface dans l'intrusion granitique Climax. Après le développement du site, 11 montages du "spent fuel" ont été mis en place pour l'épreuve en avril et en mai 1980. En même temps, 6 simulateurs électriques et 20 appareils de chauffage à garde ont été énergisés pour commencer la période d'épreuve à haute température.

Les données qui concernent le transfert de chaleur, les réponses thermomécaniques, les dosages de radiation, et les dommages de radiation ont été obtenus, et sont présentés ici, au besoin, avec des resultats calculés. En général, les resultats mesurés et calculés s'accordent bien.

*Work performed under the auspices of the U.S. Department of Energy by Lawrence Livermore National Laboratory under Contract W-7405-Eng-48.

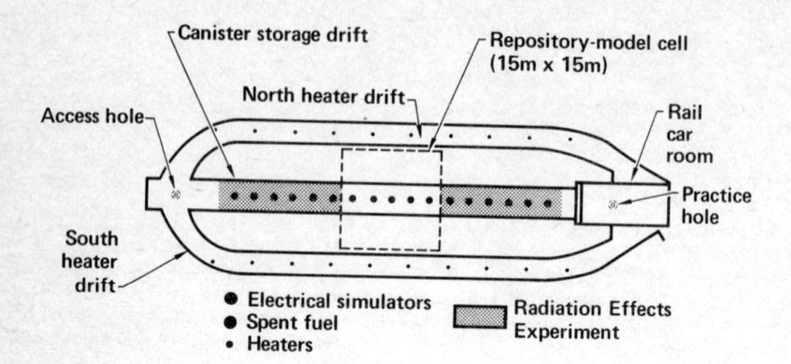

Figure 1 Plan view of the Spent Fuel Test-Climax showing repository model cell and radiation effects experiment.

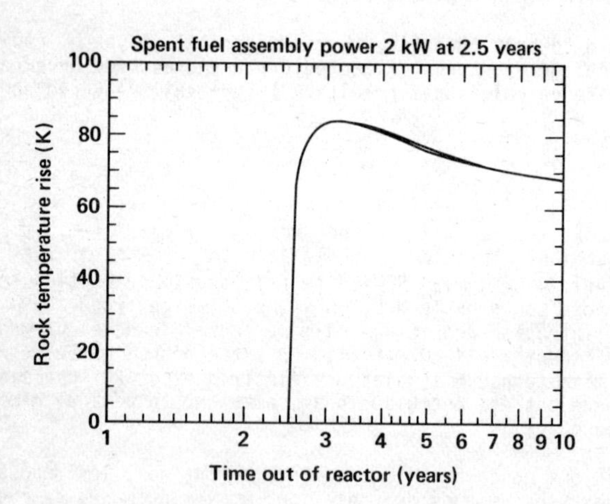

Figure 2 Comparison of calculated temperature rises on the center canister of an 8,000 canister array and the center canister of the Spent Fuel Test-Climax.

INTRODUCTION

The Spent Fuel Test-Climax (SFT-C) is being conducted under the technical direction of the Lawrence Livermore National Laboratory for the U. S. Department of Energy (DOE) at the DOE Nevada Test Site (NTS). Authorization for the test was received in June 1978. An intensive program of site characterization, facility construction, and handling system fabrication culminated in the emplacement of 11 spent fuel assemblies 420 m below surface in the Climax granitic stock during April-May 1980.

We discuss here key test results obtained during the initial two years of spent fuel storage. Test objectives are presented to provide the rationale for the SFT-C. Brief descriptions of geologic characterization, facility development, and the spent fuel handling system provide a framework for discussing the test results.

TEST OBJECTIVES

The overall objective of the SFT-C is to demonstrate the safe and reliable packaging, handling, storage, and retrieval of spent nuclear reactor fuel in a deep geologic media. A number of technical objectives have been developed to address the ultimate qualification of granite as a media for nuclear waste emplacement and to provide data useful in designing a waste repository [1].

In order to meet these objectives, it was necessary to design a facility which would simulate the conditions anticipated to be present in a repository of the future. Thermal analyses showed that this simulation could be obtained using three parallel rows of thermal sources (Fig. 1). The center 5 x 5 m region of the test array experiences the same thermal regime present in a 8,000 canister array of a repository, during the first five years of storage [2] (Fig. 2).

Also shown in Fig. 1 are six electrical simulators which are intersperced with the spent fuel assemblies in the regions outside the 5 x 5 m repository model cell. This arrangement allows a comparison of the effects on granite of heat alone (where the simulators are present) with the combined effects of heat and ionizing radiation (where the spent fuel assemblies are present).

SITE CHARACTERIZATION

Site characterization relied on available data during the early phases of planning the test [3]. Later, four 76 mm (NX) diameter exploratory cores were obtained to investigate the lateral and vertical variability of the stock prior to construction. These cores and others (obtained from instrumentation and heater holes) account for a total of 1570 m of core obtained and analyzed [4].

Fracture mapping progressed with excavation and eventually resulted in the mapping of over 2500 geologic features [5]. Structurally significant features have been identified and are included in analyses of rock response to excavation [6]. Detailed analyses of site geology are in progress.

FACILITY DEVELOPMENT

Development of the SFT-C took advantage of existing facilites which included a personnel and materials shaft, hoist and headframe, and associated surface plant. A shaft was bored 0.76 m diameter by 420 m deep to provide for lowering the encapsulated spent fuel from the surface to the underground workings [7].

Underground construction included driving two 3.4 x 3.4 m heater drifts and a 4.6 x 6.1 m high canister storage drift. Seventeen 0.61 m diameter by 5.2 m deep canister emplacement holes were drilled vertically downward on 3 m centers to accommodate the 11 spent fuel canisters and 6 electrical simulators. An eighteenth hole was provided for practice fuel-handling operations. These holes are lined with 0.46 m diameter carbon steel liners.

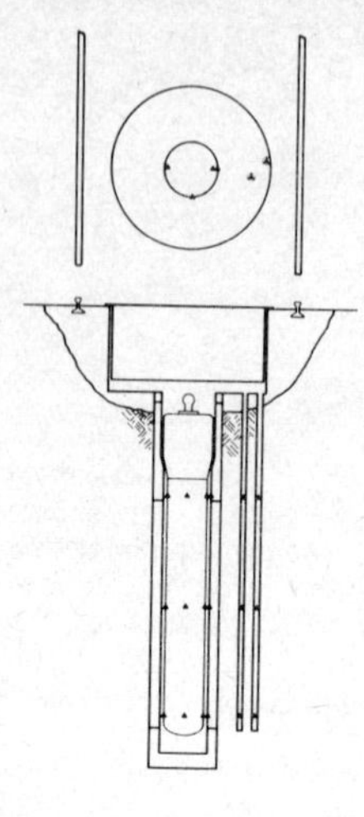

Figure 3 Storage hole geometry showing near-field thermocouple locations.

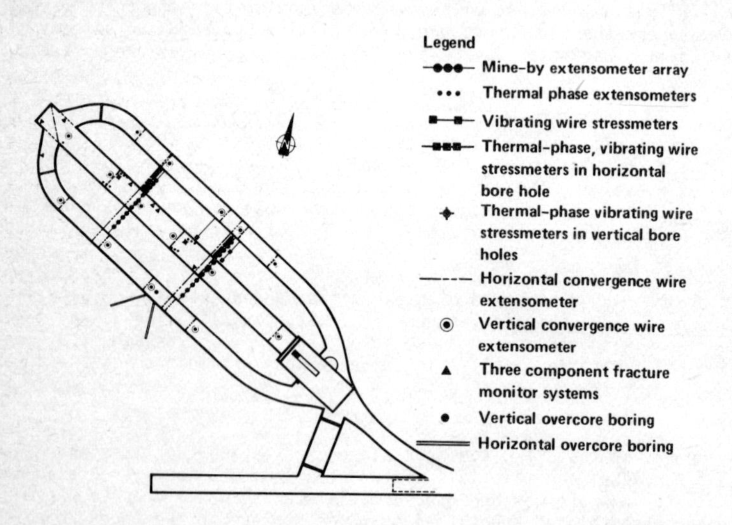

Figure 4 Plan view of Spent Fuel Test-Climax showing locations of selected geotechnical instrumentation.

An array of nearly 1000 instruments have been installed to measure temperatures, displacements, stresses, air quality, radiation dose to granite and personnel, and acoustic emissions (Figs. 3 and 4). Details of the instrumentation plan are provided by Brough and Patrick [8]. The acquisition of test data from most of these instruments is performed by a dual disc-based HP-1000 mini-computer system [9].

SPENT FUEL HANDLING SYSTEM

A three-component system was developed to facilitate transport of the spent fuel assemblies on the NTS. The spent fuel was shipped from the Turkey Point #3 (Florida) pressurized water reactor to NTS via licensed transporter. Subsequent to encapsulation at the Engine Maintenance, Assembly, and Disassembly (EMAD) facility in southwest NTS, the canisters are moved to the SFT-C in a shielded surface transport vehicle (STV). The canister lowering system (CLS) lowers the canister through the previously described drilled shaft to the underground transfer vehicle (UTV). The UTV in turn transfers the canister to and emplaces it in one of the 17 storage holes [10]. The process is reversed for retrieval operations.

RESULTS OF SPENT FUEL HANDLING OPERATIONS

The 11 encapsulated spent fuel assemblies were placed in temporary storage at the SFT-C during April and May 1980. Exchanges of spent fuel canisters between the SFT-C and EMAD are conducted on a regular basis to insure that personnel and handling systems are qualified and operational. Three such exchanges--in January 1981, October 1981, and August 1982-- have been conducted to date. Handling systems are heavily shielded and no significant radiation exposures have occurred to operating personnel.

RESULTS OF TECHNICAL MEASUREMENTS PROGRAM

An extensive field measurements program provides data to address test objectives. Laboratory studies provide data on critical phenomena which may otherwise be difficult to interpret due to the high variability of in situ conditions.

Instrumentation and Data Acquisition

The reliable performance of the data acquisition system and instrumentation are essential to meeting the technical objectives of the test and are therefore evaluated here. The HP-1000 based data acquisition system which records data from nearly 1000 instruments is functioning reliably; the monthly average of the functionally disabled index (FDI) has not exceeded ~10% since May 1980. This high degree of reliability is a direct result of utilizing two computers which augment each other's performance -- the FDI for an individual computer often exceeded 25%. System accuracy has been within a ± 4 microvolt and a ± 9.2 milliohm window for DC voltage and resistance, respectively [11]. Over 6 million data records have been acquired for analysis.

Instrumentation has also functioned reliably with two exceptions. First, a group of linear potentiometers failed by becoming nonlinear in resistance. Second, nearly all of 18 vibrating-wire stressmeters failed due to internal corrosion [12]. Replacement of these transducers has occurred and instrument evaluation continues.

Thermal Transport

Two- and three-dimensional thermal transport calculations were performed using the TRUMP finite difference code [2]. The calculational mesh is 80 x 160 m and contains over 2200 zones. The time constant for this geometry is ~120 years.

The thermal properties of rock selected for use in the code were initially based on laboratory measurements and were later revised to reflect data

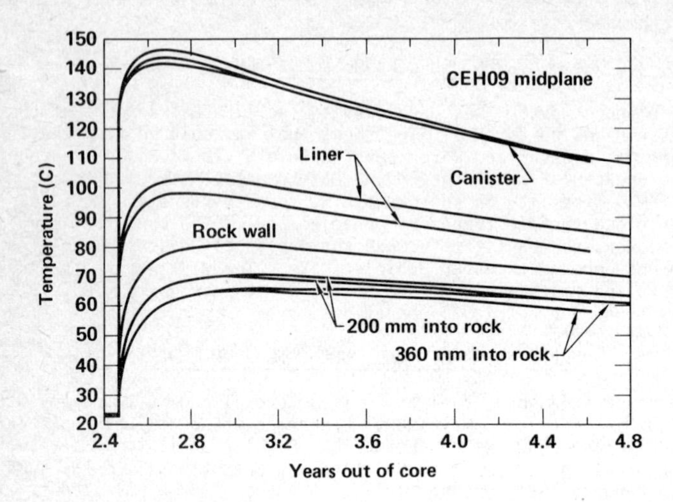

Figure 5 Temperature histories at selected radial locations at the axial midplane of the center spent fuel assembly.

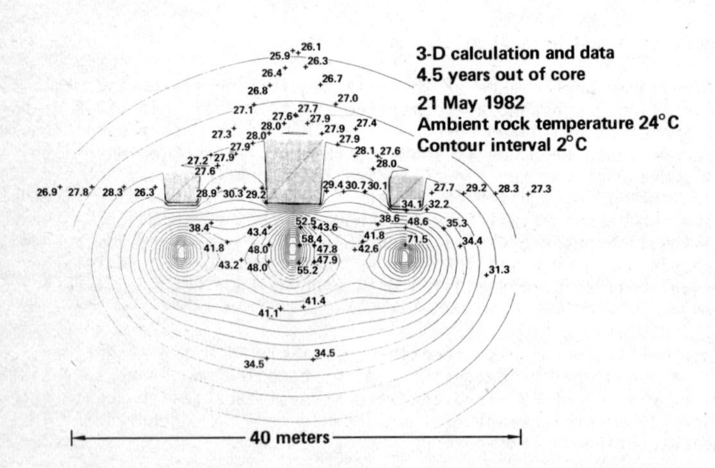

Figure 6 Calculated temperature contours and measured temperatures in a cross-sectional plane of the Spent Fuel Test-Climax after 2 years of heating.

obtained from a small-scale heater test [13]. Emittance and convection coefficient data were not obtained for this test so values reported in the literature were used.

The effects of ventilation were simulated using a partial flow model which was developed specifically for the SFT-C calculation.

Electrical energy input to the facility is measured with watt transducers and is thus both controllable and known. Energy input from the spent fuel was initially calculated, to determine the shape of the power-time curve, and then normalized to an observed calorimetry result [14].

Near- and intermediate-field temperature measurements show that conditions in the repository model cell closely approximate those calculated (Figs. 5 and 6). This indicates a high level of success in modeling heat transfer in the rock mass. We are currently investigating possible sources of the 1-5°C differences observed [11].

Ventilation and dewpoint measurements indicate that about 13.3% (75 MW•h) of the input energy to the SFT-C was removed in the ventilation airstream; as compared to about 1/3 calculated. Of this total, 74% was removed as sensible heat and 26% was removed as latent heat of vaporization. The latter is the result of removing 31,000 L of water from the facility in the airstream during the first 1 1/2 years of spent-fuel storage. Measured heat removal rates vary significantly from those calculated (Fig. 7), indicating that we are inadequately modeling the complexities of heat transfer into the airstream and out of the facility through the ventilation system [11].

Thermomechanical Response

Thermomechanical response calculations were performed using ADINA and the compatible ADINAT heat-flow code [15]. Thermal properties used here were the same as used in the thermal calculations with one exception. Since ADINAT is unable to treat either internal radiative heat transfer or ventilation effects, "equivalent" material properties had to be developed to simulate these phenomena. A comparison of ADINAT and TRUMP results was used to verify the correct choice of properties [16].

Elastic properties of the rock were initially obtained from laboratory testing. *In situ* measurements of Young's modulus and Poisson's ratio provided more realistic rock mass properties and also indicated the presence of a 0.5 m thick zone of low-modulus rock surrounding the three drifts [17]. This zone is also modeled in the calculations.

In accordance with measurements [18], the stress boundary conditions on the calculations are gravitational in the vertical direction and 1.2 times gravitation in the horizontal direction.

Agreement between measured and calculated displacements has been quite good [19]. Displacements in most locations have not exceeded 2 mm, as measured over distances of 3.5 to 6 m (Figs. 8 and 9). The good agreement observed to date implies that geologic features are not as significant during the heated phase of the experiment as they were during the mine-by experiment [6].

Acoustic emission (AE) monitoring records the occurrence of small-scale fracturing in the granite. These AE "events" are typically the result of stress-induced fractures a few centimetres in length. Background rates of 2-3 events per week were recorded. This rate increased dramatically to 15-20 events per week shortly after the spent fuel was emplaced and the heaters were energized (Fig. 10). With the exception of two short periods of activity in excess of 10 events per week, rates have returned to a low of a few events per week [11]. As expected, no instability of boreholes or underground openings has occurred as a result of this small-scale fracturing.

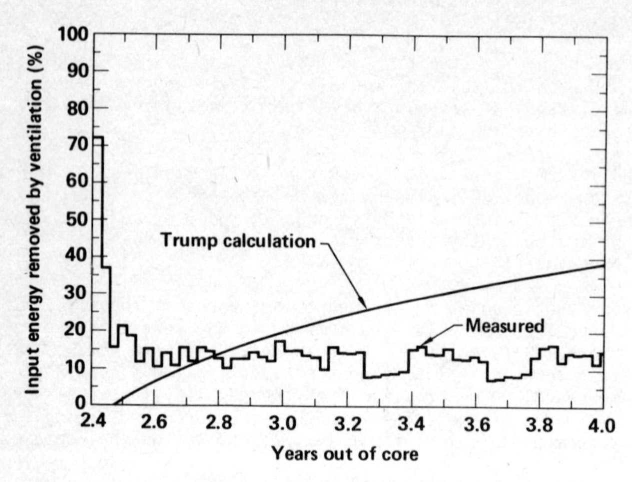

Figure 7 Calculated and measured heat removal rates at the Spent Fuel Test-Climax.

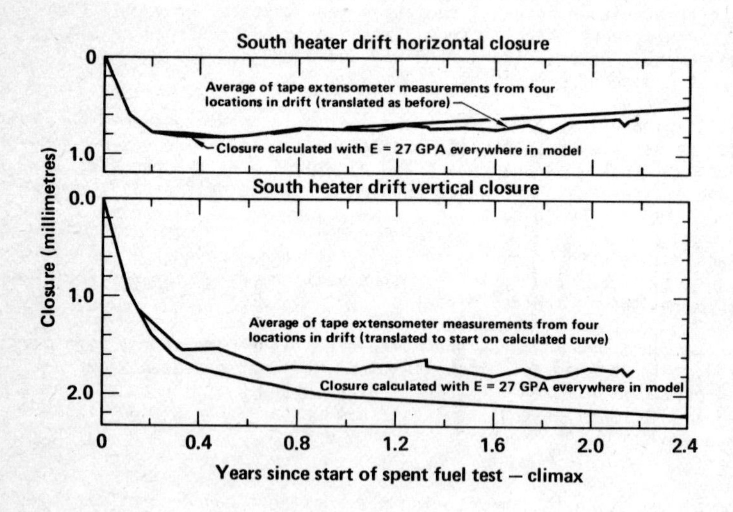

Figure 8 Calculated and measured closures occurring in the south heater drift of the Spent Fuel Test-Climax.

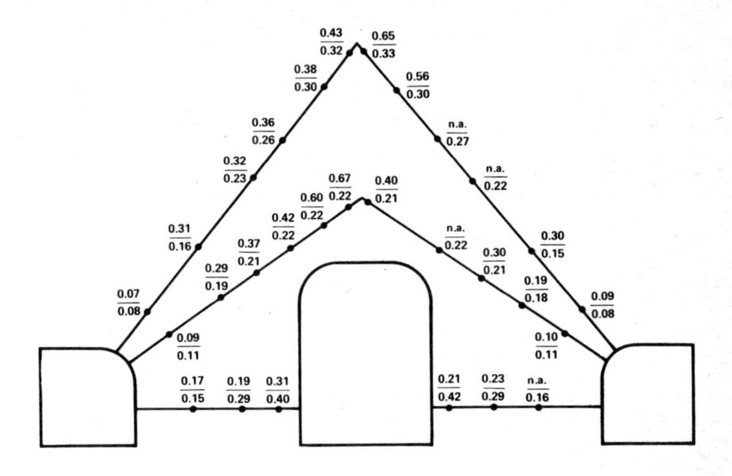

Figure 9 Calculated and measured displacements occurring in a cross-sectional plane of the Spent Fuel Test-Climax after 1 1/2 years of heating.

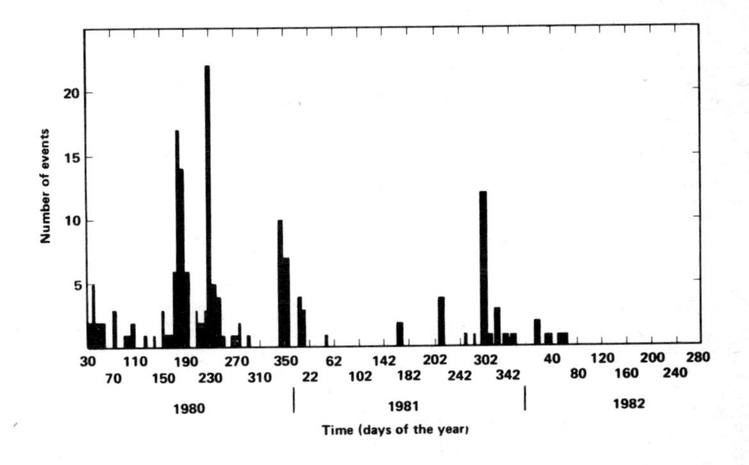

Figure 10 Summary of acoustic emission activity at the Spent Fuel Test-Climax.

Radiation Effects and Transport

Laboratory investigations of Climax granite indicate a statistically significant degradation in uniaxial compressive strength (~20%) and Young's modulus (5-10%) after gamma irradiation at a total dose of 13 MGy. Brazilian tensile strength of the same material was not measurably affected by the same treatment [20].

Radiation transport calculations were performed using an in-house Monte Carlo radiation transport code named MORSE-L [21]. These calculations were performed to provide radiation dose-to-granite data for use in assessing the relative effects of heat alone (where electrical simulators were deployed) versus heat and radiation (where spent-fuel assemblies are stored). The storage-hole geometry was modeled explicitly in two dimensions. Calculations showed a maximum integrated dose to granite of 1.6 MGy for a three-year storage phase.

Radiation dose measurements are made at three spent-fuel emplacement holes using LiF dosimeters. In addition, a set of pre-irradiated dosimeters are installed on an electrical simulator to examine the effects of annealing which occur as a result of time and elevated temperature.

Radiation-dose-to-granite measurements have indicated several problems in current measurement technology in a high dose, elevated temperature environment. Annealing and fading effects during and after irradiation lead to complex calibrations for the dosimeters and high levels of variability in the data [22]. Due to this variability, comparison with pre-test dose calculations has not been meaningful to date.

FUTURE ACTIVITIES

The SFT-C was originally planned to be a 3-5 year test. The technical measurements support a storage phase duration of three years, since essentially all of the thermal and thermomechanical response of the rock will have been recorded by that time. We therefore plan to retrieve the spent fuel and return it in laq storage at EMAD in FY 1983.

We will continue to monitor the response of the facility during an approximately six-month cool-down period. Post-test sampling and analysis of geologic and man-made materials will follow. Field activities will be completed by the end of FY 1984 and final test documentation will be completed by mid-FY 1985. Plans for the future utilization of the SFT-C facility are being considered.

REFERENCES

1. Ramspott, L. D., et al.: Technical Concept for a Test of Geologic Storage of Spent Reactor Fuel in the Climax Granite, Nevada Test Site, Lawrence Livermore National Laboratory, UCRL-52796, June 1979.

2. Montan, D. and W. Patrick: Thermal Calculations for the Design, Construction, Operation, and Evaluation of the Spent Fuel Test-Climax, Nevada Test Site", Lawrence Livermore National Laboratory, URCL-53251, September 1981.

3. Maldonado, F. : Summary of the Geology and Physical Properties of the Climax Stock, Nevada Test Site, U.S. Geological Survey Open File Report 77-356, 1977.

4. Wilder, D., J. Yow, R. Thorpe: Core Logging for Site Investigation and Instrumentation, Lawrence Livermore National Laboratory, UCRL-in preparation

5. Wilder, D. and J. Yow: Fracture Mapping at the Spent Fuel Test-Climax, Lawrence Livermore National Laboratory, UCRL-53201, May 1981.

6. Heuze, F., T. Butkovich, J. Peterson: An Analysis of the 'Mine-by' Experiment, Climax Granite, Nevada Test Site, Lawrence Livermore National Laboratory, UCRL-53133, June 1982.

7. Patrick, W. C. and M. C. Mayr: Excavation and Drilling Activities Associated with a Spent Fuel Test Facility in Granitic Rock, Lawrence Livermore National Laboratory, UCRL-53227, November 1981.

8. Brough, W., W. Patrick: Instrumentation Report #1: Specification, Design, Calibration, and Installation of Instrumentation for an Experimental, High Level, Nuclear Waste Storage Facility, Lawrence Livermore National Laboratory, UCRL-53248

9. Nyholm, R., W. Brough and N. Rector: Data Acquisition System Integration Report, Lawrence Livermore National Laboratory, UCRL-in preparation, 1982.

10. Duncan, J. E., P. A. House, and G. W. Wright: Spent Fuel Handling System for a Geologic Storage Test at the Nevada Test Site, Lawrence Livermore National Laboratory, UCRL-83728, May 1980.

11. Patrick, W. C. et al.: Spent Fuel Test-Climax: Technical Measurements Interim Report FY 1981, Lawrence Livermore National Laboratory, UCRL-53294, April 1982.

12. Patrick, W. C., R.C. Carlson, and N. L. Rector: Instrumentation Report #2: Identification, Evaluation, and Remedial Actions Related to Transducer Failures at the Spent Fuel Test-Climax, Lawrence Livermore National Laboratory UCRL-53251, November 1981.

13. Montan, D. and W. Bradkin: Granite Heater Test No. 1, Lawrence Livermore National Laboratory, in preparation.

14. Schmittroth, F., G. J. Neely and J. C. Krogness: A Comparison of Measured and Calculated Decay Heat for Spent Fuel Near 2.5 Years Cooling Time, Hanford Engineering Development Laboratory, Richland, WA., HEDL-TC-1759 (1980), preliminary report, controlled distribution.

15. Butkovich, T.: Thermomechanical Response Calculations of the SFT-C in As-Built Configuration, Lawrence Livermore National Laboratory, UCRL-53198, June 1981.

16. Butkovich, T. and D. Montan: A Method for Calculating Internal Radiation and Ventilation with the ADINAT Heat-Flow Code, Lawrence Livermore National Laboratory, UCRL-52918, April 1980.

17. Heuze, F., W. Patrick, R. De la Cruz, and C. Voss,: In Situ Geomechanics, Climax Granite, Nevada Test Site, Lawrence Livermore National Laboratory, UCRL-53064, December 1980.

18. Ellis, W. and J. Magner,: Determination of In Situ Stress in Spent Fuel Test Facility, Climax Stock, Nevada Test Site, Nevada, U.S. Geological Survey Open File 82-458.

19. Yow, J. L., Jr., and T. R. Butkovich: Calculated and Measured Drift Closure During the Spent Fuel Test in Climax Granite, Lawrence Livermore National Laboratory, UCRL-87179, April 1982.

20. Durham, W. B.: The Effect of Gamma Irradiation on the Strength of Climax Stock Granite, Lawrence Livermore National Laboratory, UCRL-87475, March 1982.

21. Radiation Dose Calculations for Geologic Media Around Spent Fuel Emplacement Holes in the Climax Granite, Nevada Test Site, Lawrence Livermore National Laboratory, UCRL-53201, May 1981.

22. Quam, W. and T. DeVore: Climax Spent Fuel Dosimetry Progress Report, September 1980-September 1981, Lawrence Livermore National Laboratory Contractor Report UCRL-15419, EG&G 1183-2432, 1981.

RESULTS AND CONCLUSIONS FROM ROCK MECHANICS/HYDROLOGY INVESTIGATIONS: CSM/ONWI TEST SITE

W. Hustrulid
Colorado School of Mines
Golden, Colorado, U.S.A.

W. Ubbes
Office of Nuclear Waste Isolation
Battelle Memorial Laboratories
Columbus, Ohio, U.S.A.

ABSTRACT

The Colorado School of Mines (CSM) under sponsorship of the Office of Nuclear Waste Isolation (ONWI) has established a hard rock test facility in its Experimental Mine at Idaho Springs, Colorado. The CSM/ONWI Experimental Room (30m long, 5m wide and 3m high) was driven in the jointed biotitic gneiss using carefully designed and executed blast rounds. This room is above the water table, and lies under approximately 100m of cover. Included in this paper are:

- the results of RQD, crosshole ultrasonic, permeability and deformation modulus measurements made as a function of distance away from and position around the perimeter of the room (location of Round 4). These measurements have been made for the purpose of evaluating the extent and degree of excavation disturbance.

- the results of a series of packer tests conducted in one 33m long borehole presented in the form of a probability-permeability curve of the Weibull type. Experimental results regarding the effect of interval length on observed permeability are discussed.

- an overview of the instruments selected and measurements used by Terra Tek in determining thermal and mechanical properties from a heated block experiment. This block test was conducted by Terra Tek under ONWI sponsorship in the Experimental Room.

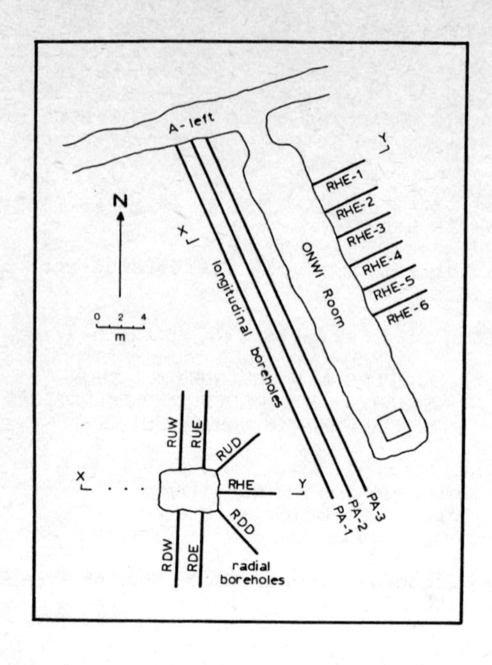

Figure 1. Plan Layout of the CSM Experimental Mine

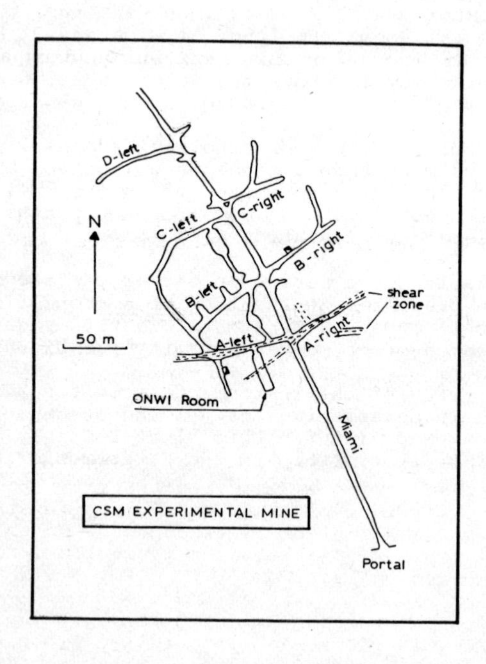

Figure 2. Diagrammatic Representation of the CSM/ONWI Hard Rock Facility

1. INTRODUCTION

The excavation of a hard rock test facility funded by the Office of Nuclear Waste Isolation (ONWI) under contract to the Department of Energy (DOE) was begun at the CSM Experimental Mine, Idaho Springs, Colorado in July 1979. Three aspects of the project were

- demonstration and evaluation of careful excavation techniques in hard rock
- characterization of the disturbance zone around the excavation
- creation of an underground research laboratory in which a wide variety of geomechanics experiments could be performed.

The expected products from the program include

- procedures/techniques/instruments that could be applied at other hard rock sites
- an extensive data base for modelling efforts
- a well characterized facility for use in R/D efforts by other researchers.

It is impossible to adequately present the results and conclusions from the work done to date in this short paper. The titles and status of some of the reports prepared on the site are included in Appendix A. Rather an attempt will be made to highlight some of the findings that could have an immediate impact on the planning of experiments being done by others. In particular the paper will address

- characterization of the rock surrounding an underground opening
- effect of scale on permeabilty measurements
- instrumentation for heated block tests in jointed rock

The discussion of the first two topics is based upon work done by CSM (1, 2, 3). The last topic describes results obtained by Terra Tek (4).

2. DESCRIPTION OF THE CSM/ONWI FACILITY

Figure 1 shows the plan view of the mine at the 2400m (7880 ft) level. The CSM/ONWI room is approximately 100m (300 ft) below the ground surface and is 20m (100 ft) long, 3m (10 ft) high and 5m (15 ft) wide.

Forty-five NX diameter (78.5mm) boreholes were diamond drilled in and around the room shortly after its excavation. Forty-two of these were drilled from inside the room in six radial sites. The seven holes of each set are arranged as shown in Figure 2. The west side of the room is paralleled by three 33m long NX diameter boreholes drilled from A-left.

Terra Tek excavated a 2mx2mx2m block in the floor at the far end of the room. This block was line drilled along the sides with the bottom remaining intact (Figure 3). Eight sandwich pairs of flatjacks (2m x 1m) grouted into the slots were used to generate uniaxial or biaxial stresses in the range of 0 to 7 MPa. Nine 1 kw heaters were located along the midplane of the block.

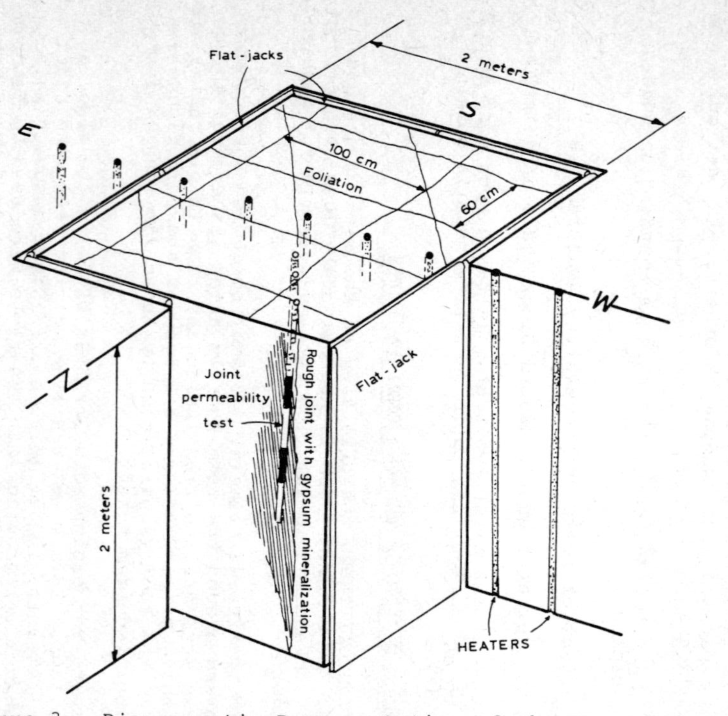

Figure 3. Diagrammatic Representation of the Terra Tek Heated Block Test. Reference (4)

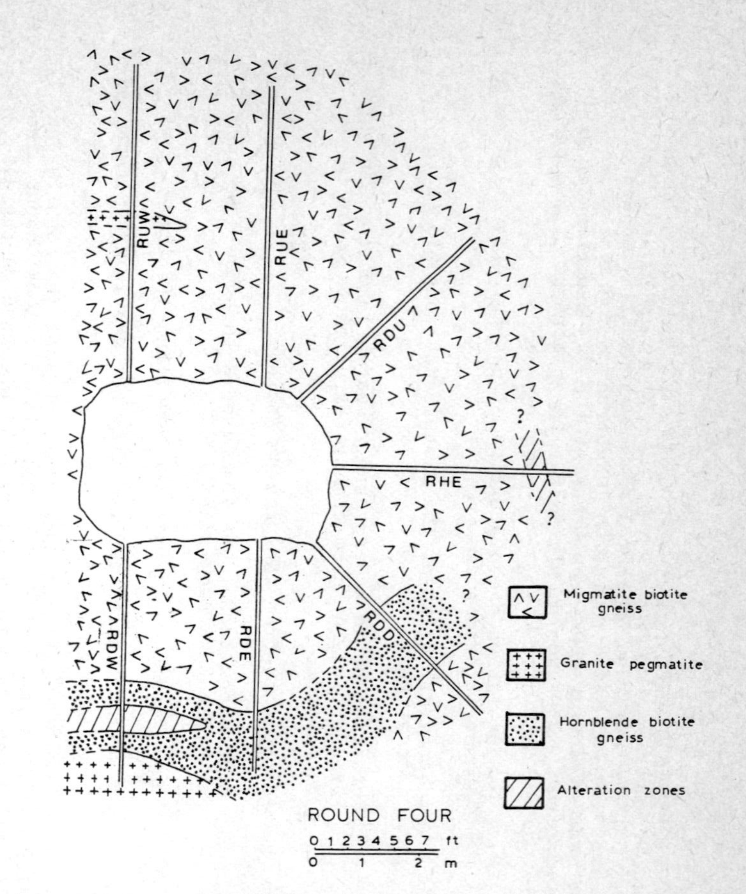

Figure 4. Geologic Section Through Round Four. Reference (3)

The rock around the room is strongly foliated with the foliation striking N70E and dipping 70NW. The main rock type is a medium to coarse-grained migmatite biotite gneiss.

Two major vertical fracture sets are present, striking N50-60E and N40-50W. The first set is parallel and the second is nearly perpendicular to the foliation.

3. CHARACTERIZATION OF THE ROCK IN THE VICINITY OF AN EXCAVATION

3.1 Introduction

The room was excavated using blast rounds that were designed to minimize the extent of damage to the surrounding rock. The degree and extent of damage has important ramifications for (a) excavation method selection, (b) evaluating sealing requirements/possibilities, and (c) modelling thermal/mechanical/hydrologic response. One could visualize the final "disturbance zone" as due to a superposition of blast damage and stress redistribution. Due to blast damage alone the zone around opening would be expected to exhibit

- lower modulus (E)
- higher permeability (K)
- lower P and S wave velocities (V_p and V_s respectively)
- lower Rock Quality Designation (RQD)

than rock not affected by the blasting. With the removal of the excavated rock the stresses that were previously transmitted through the opening are transferred to the surrounding rock. The result is a change in the stress state surrounding the opening. Since all of the above with the exception of RQD are stress dependent with an increase in stress one would expect

- higher modulus
- lower permeability
- higher wave velocities

The favorable superposition of the material/stress change conditions could lead to an apparent reduction in

- the degree of damage
- the extent of the damage zone.

An unfavorable superposition could accentuate the degree and extent of damage. A whole series of techniques/procedures have been used in the radial boreholes to try and separate the two contributions to the apparent blast damage zone. The results have not as yet been fully analyzed and any conclusions would be premature. Some results for Round 4 and comments regarding stress measurements in jointed rock will be made.

3.2 Some Results For Round 4

Figure 4 is a vertical geologic section drawn through Round 4.

The geotechnical characterization of the rock in the vicinity of the opening has been made using a number of techniques including amongst others

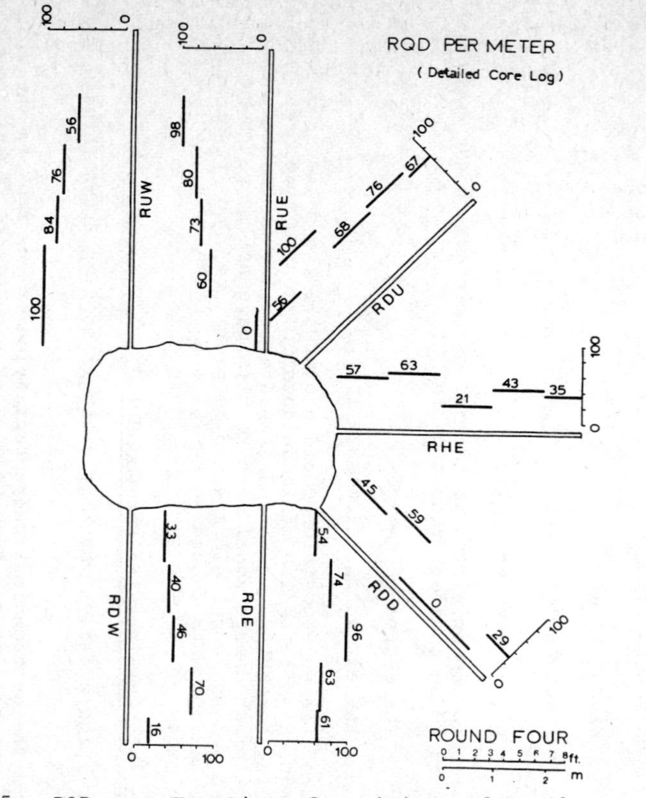

Figure 5. RQD as a Function of Position and Depth for Round Four. After Reference (3)

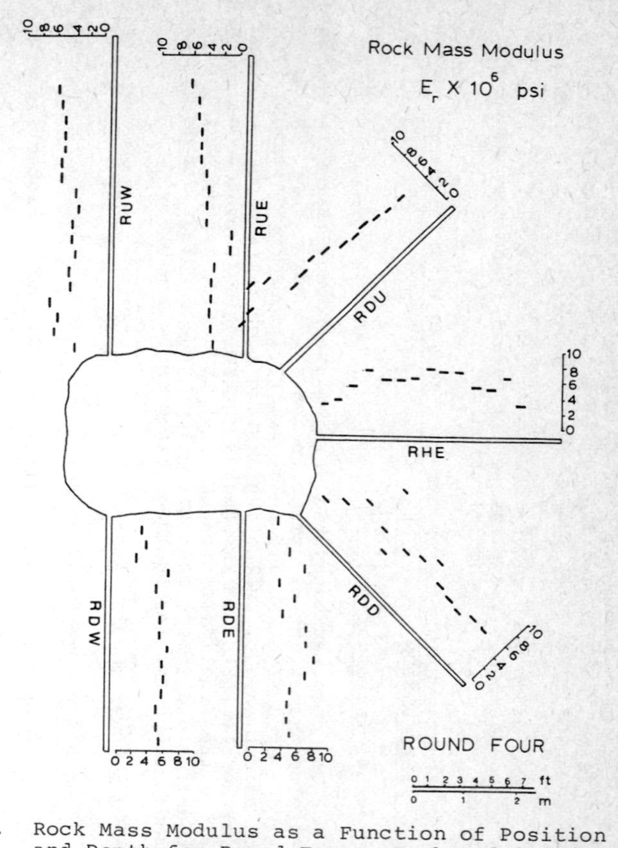

Figure 6. Rock Mass Modulus as a Function of Position and Depth for Round Four. Rock Modulus (GPa) = 6.9 x Rock Modulus (10^6 psi). After Reference (2)

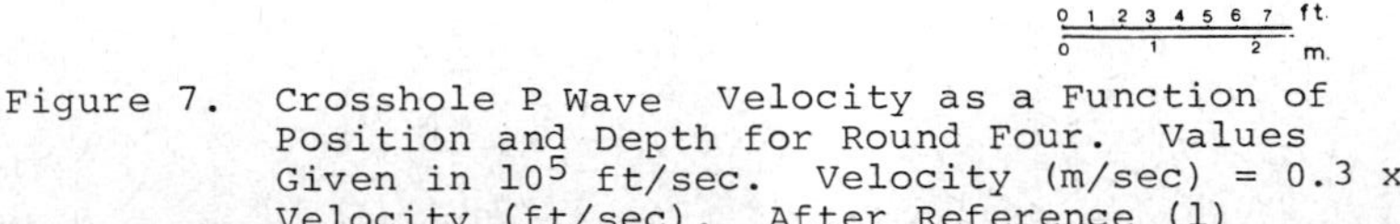

Figure 7. Crosshole P Wave Velocity as a Function of Position and Depth for Round Four. Values Given in 10^5 ft/sec. Velocity (m/sec) = 0.3 x Velocity (ft/sec). After Reference (1)

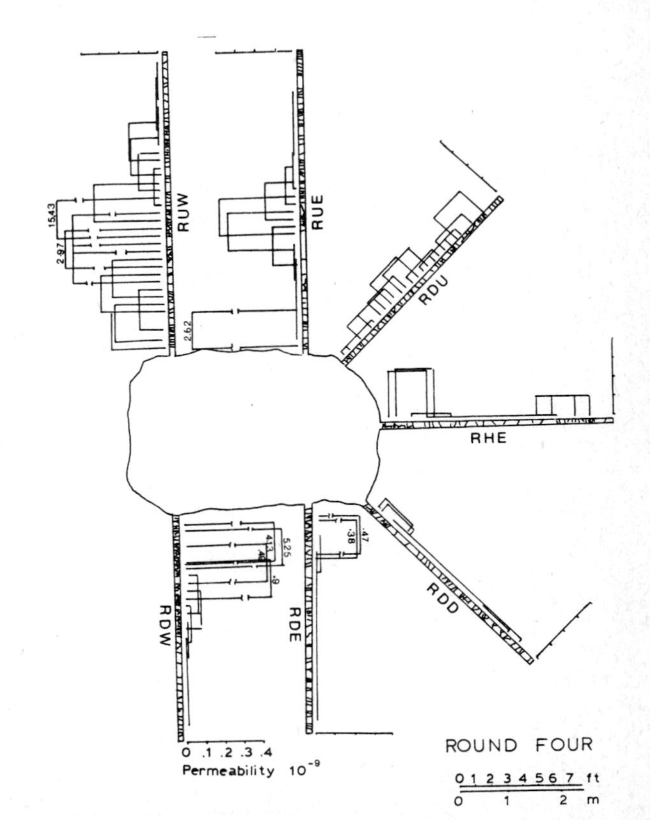

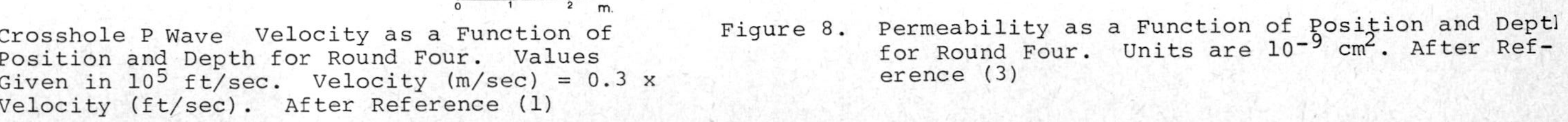

Figure 8. Permeability as a Function of Position and Deptl for Round Four. Units are 10^{-9} cm^2. After Reference (3)

· Rock Quality Designation (RQD)

· Modulus (E) determinations using the CSM cell

· Crosshole ultrasonic measurements (V_p)

· Double packer permeabilities (DPP)

The detailed results for Round 4 are given in Figures 5 through 8. A comparison of the apparent blast damage boundaries as measured using the different techniques is given in Table I.

Table I. Comparison of the Extent of Disturbance Around the Room Using Four Techniques Depth(m) = 0.3 x Depth (ft.)

Position	Depth of Disturbance Zone (ft) by Technique			
	RQD	E	Vp	DDP
RUW	0	2	-	2
RUE	3	2	1	3
RDU	3	1	1	0
RHE	3	3	3	3
RDD	3	3	1	3
RDE	3	3	2	3
RDW	3	2	-	4

This distance is of the same order of magnitude predicted using formulas developed by Holmberg (5). A great deal of analysis remains to be done concerning the meaning of the individual results as well as the relationships between them.

Some conclusions regarding the instruments/techniques used are

· the NX version of the CSM cell is a simple and reliable borehole tool that provides reproducible data

· the success of the crosshole ultrasonic system is highly de-pendent on the type and number of discontinuities present. Transmission distance using 200 kHz frequency crystals was of the order of a meter. For a crystal frequency of 80 kHz it was sometimes possible to transmit across 5m.

· the double packer permeability equipment yielded good results and reproducibility.

3.3 Comments on Stress Measurements in Jointed Rocks

Three overcoring techniques (USBM Borehole Deformation Gage, CSIRO Hollow Inclusion Stress Meter, Lulea (Modified Leeman) Triaxial Cell) were used in four horizontal parallel holes drilled into the West wall of the room for the purpose of determining the near and far field stresses. The results have been, to say the least, extremely difficult to evaluate. For the interpretation of results from all three techniques one depends upon linear elasticity and a knowledge of the appropriate elastic constants. Most data reduction routines also require isotropy. If testable samples are not available from the overcoring locations, then one applies an assumption of homogeniety. Problems in interpretation have been experienced due to

· jointed nature of the rock mass

· foliation

· anisotropy

• non-homogeniety

• questions regarding the performance of instruments

• linearity of overcore sample vs. nonlinearity of the jointed mass containing the sample.

Major questions remain regarding the meaning of measured deformations/strains in terms of the local stress field and the use of such data in the interpretation of an overall "average" stress field.

4. PERMEABILITY MODELLING AND SCALE EFFECTS

4.1 Introduction

An extensive permeability measurement program was carried out at the facility. Since the rock is considered to be unsaturated, special testing techniques were required. The present authors have selected one set of results to illustrate one possible way of presenting permeability data in a form that might be useful for site characterization. There are certainly many other ways and the method selected may not be the best (or even appropriate). It is presented here to focus attention on the importance of result format as well as the results themselves. The permeabilities determined at various positions along the 32m long borehole PA-2 are shown in Figure 9. The interval pressurized with nitrogen has a length of 7 feet. As can be seen, there is quire a high variation along the hole with the maximum permeability occuring at a distance of 20m in a shear zone. It is also of interest to examine the distribution of permeabilities obtained with the 7-ft. packer interval. The permeability values falling in the range 2.7×10^{-13} to 4×10^{-11} cm^2 were used to determine the equation given below

$$S = 100 \left[1 - e^{-\left(\frac{K}{1.7097}\right)^{0.8133}} \right]$$

where

S = cumulative probability (%) of permeability K or smaller

K = permeability (10^{-11} cm^2)

This equation has been superimposed upon the data points in Figure 10. As can be seen, the fit is very good for the range selected. The equation would suggest that no interval would be encountered of permeability greater than about 10^{-9} cm^2. The probability for the occurrence of any particular permeability is also given. Unfortunately the higher permeability data depart from the predicted curve suggesting that these zones may contain fractures which belong to a different class of discontinuities. The conclusions to be drawn are (a) the probability-permeability curve is of an interesting form for application, (b) it would appear to help identify different fracture groups, (c) one curve does not fit the entire range of permeabilities encountered.

A series of tests were run in hole PA-2 in which the bottom of the hole formed one packer and the permeability was determined as the other packer was moved toward the hole collar. The results are given in Table II. Also presented are the maximum permeabilities observed for that interval using the 7-ft. interval chamber. As can be seen the presence of the high permeability zone (4.5×10^{-10} cm^2) influences the measured overall permeability of the zone very little. The maximum value is about a factor of 5 times higher than the average detected with the long interval suggesting that "average" or "overall" permeabilities should be applied with care.

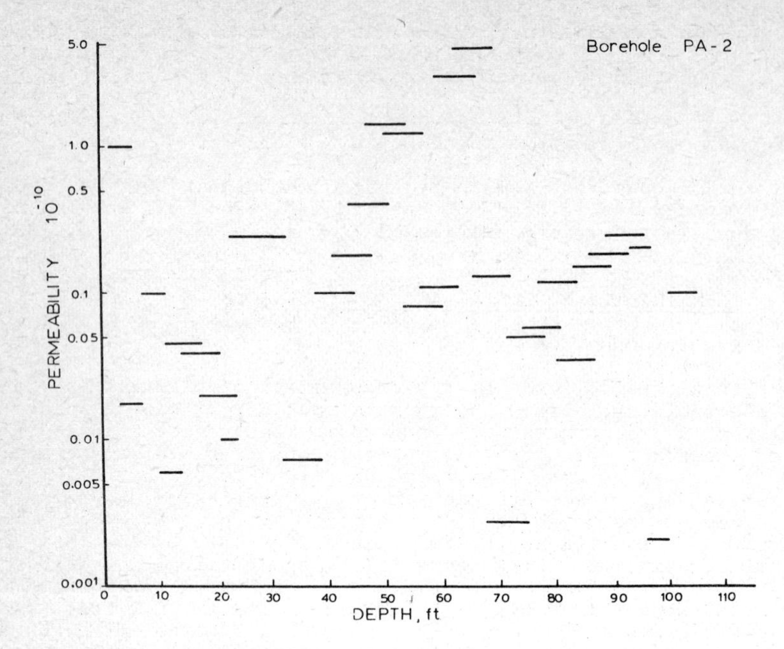

Figure 9. Permeability as a Function at Distance Along a Borehole. Depth (m) = 0.3 x Depth (ft). After Reference (3)

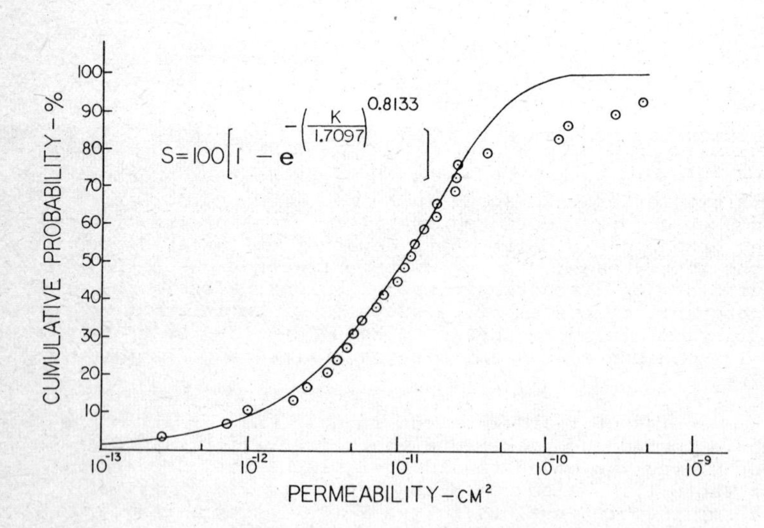

Figure 10. Cumulative Probability-Permeability Curve For the PA-2 Borehole. A Curve Based Upon the Weibull Distribution Has Been Superimposed.

Table II. Measured Overall Permeabilities and Maximum Permeabilities As A Function of Interval Length (After Montazer (3))

Interval (m)	Permeabilities (10^{-11} cm^2)	
	Overall	Maximum
3.3 - 32.4	6.0	45
12.4 - 32.4	9.5	45
17.0 - 32.4	9.2	45
21.5 - 32.4	3.5	1.2
27.0 - 32.4	4.0	1.5
30.3 - 32.4	2.6	1.0
31.8 - 32.4	0.24	1.0

5. TERRA TEK HEATED BLOCK TEST

5.1 Introduction

Terra Tek under contract to ONWI designed and performed a heated block test in the jointed floor rock at the end of the CSM/ONWI Experimental Room. The overall objective of the test was to develop and evaluate a suitable test method for quantifying the mechanical and thermal behavior of a jointed rock mass in situ. Of particular interest were the following:

- measurement of stress-strain behavior at ambient and elevated temperature
- measurement of the normal and shear deformation of selected joints under applied mechanical stresses and imposed thermal strains
- measurement of the thermal fields and the resulting strains and stresses as a function of confining pressure and temperature

In addition, the dependence of ultrasonic compressive and shear wave velocity, joint permeability, and induced stresses were investigated as a function of temperature and confining pressure. The detailed results of this study are described in reference (4). In this section, the present authors have condensed portions of this report focusing on the instrumentation plan selected for the block and the types of measurements used to determine rock properties. Much of the material presented is paraphrased or reproduced directly from the Terra Tek report. It is convenient to divide the discussion into surface and subsurface instrumentation.

5.2 Surface Instrumentation Layout/Results

Layout

All surface instruments were of the same general type in that they all measured deformation/strain at or near the surface of the block. Surface instrumentation included the following gage types and quantities:

- 8 horizontal strain indicators (HSI)
- 3 bonded resistance strain gage rosettes
- 4 vibrating wire strainmeters (IRAD)
- 52 pairs of points for measurements with the Whittemore gage

Figure 11 shows the location of the Whittemore pins.

Thermal Expansion

The following sources of thermal expansion data were analyzed:

- bonded resistance strain gages
- horizontal strain indicators
- Whittemore pin measurements

The surface-bonded strain gages were used to calculate thermal strain directly. Although the surface rosettes showed a generally smooth increase in strain as a function of temperature, the magnitudes were significantly less than the other measurement techniques.

The thermal expansion data (for gage lengths of 50 to 150 cm) from the HSI horizontal strain indicators showed: 1) widely varying results and 2) considerably higher expansion coefficients than measured by other techniques. The reason for the apparently large strains is not known since the gages in general exhibited stable behavior. The strain indicators yielded the same directional sensitivity of the expansion coefficient as did the Whittemore pins.

The Whittemore pins measurements provided the best measure of horizontal expansion. It was feared that non-uniform vertical expansion of the block surface would result in a surface "bowing", thus adding an angular displacement to the Whittemore pins. Level vials attached to the horizontal strain indicator pins showed no surface tilt, thus no correction was needed.

Surface Deformation

The responses of the rock at the surface to changing load conditions was recorded successfully by two instrument types: horizontal strain indicators (HSI gages) and the Whittemore strain gages. Young's moduli were calculated for increasing and decreasing stress.

Deformation modulus was determined with most success by instruments mounted on anchor posts set 10-30 cm into the block surface. The surface measurements of block deformation show high variability due to (a) surface decoupling, (b) physical inhomogeneity of the block and (c) disturbance of the gages during necessary operations on the block during testing. The extremely high moduli suggested from some measurements were probably due to decoupling. The high variability in horizontal strain indicator (HSI gage) data was caused by the unavoidable disruption of the gages by operators engaged in Whittemore surveys of the block surface. Whittemore measurements of changes in the width of boundary cracks in the grout above each flatjack, provided estimates of moduli over the full 2 meter dimensions of the block.

Loading and unloading moduli were not tabulated for the bonded surface strain gage rosettes. These gages did respond to the applied stress, but the smaller responses yielded Young's moduli in a range from 300 to over 3000 GPa.

"Poisson" Ratios

"Poisson" ratios of transverse to axial strain were calculated using Whittemore strain data acquired during uniaxial loading at ambient temperature. Individual measurements of Poisson's ratio fell in the range of < 0 to > 2.0, probably because of shear displacement along joints.

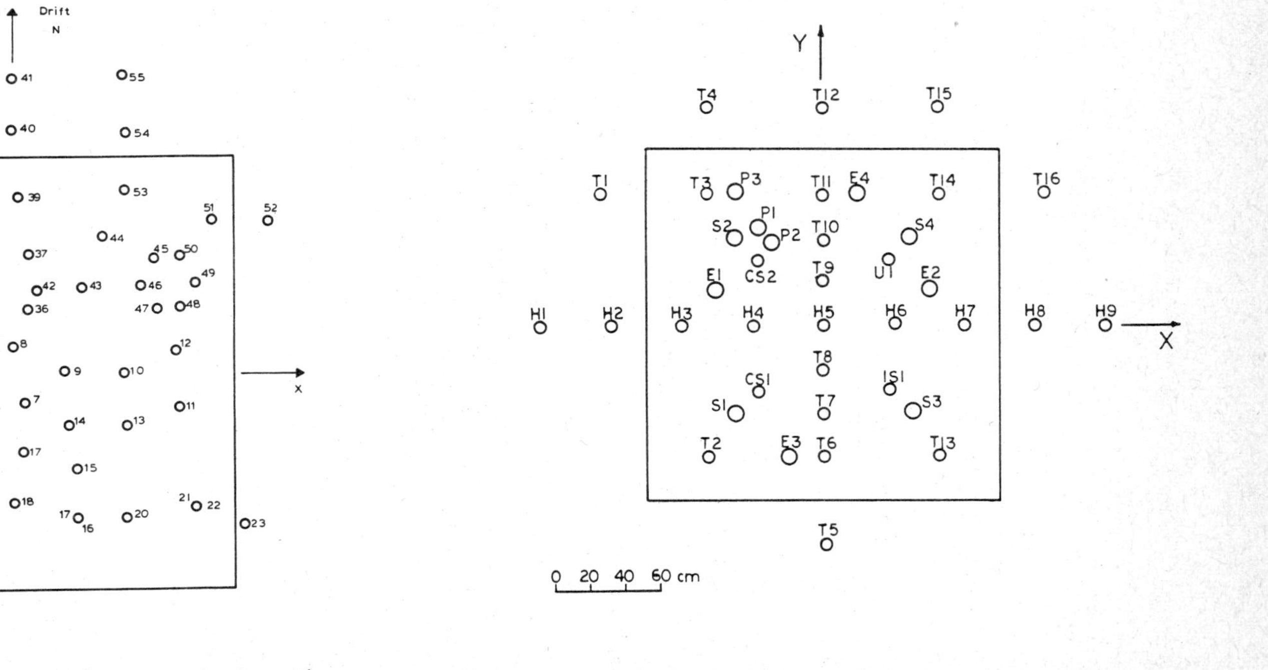

Figure 11. Position of the Whittemore Pin Locations. After Reference (4)

Figure 12. Position of the Hole Collars for Subsurface Instrumentation in the Block. After Reference (4)

Joint Normal Stiffness

Several pairs of Whittemore pins were located across visible joints intersecting the block. Several gages provided closure data for the diagonal joints, and others for the foliation joints. The majority of pins spanned what appeared from inspection to be intact rock. In each case the gage length was 10 inches (25.4 cm). The deformation of this intervening intact rock is subtracted from the overall gage response to obtain the net joint deformation. Because of the small displacements involved in normal joint closures the continuous base of the block would not be expected to affect the measurements.

Joint Shear Stiffness

The shear displacements along joints were also monitored with Whittemore pins.

The continuous base of the block means that the block will tend to behave stiffly if shear displacements along a joint become significantly greater than so-called "elastic" behavior.

The actual strength envelope would be expected to be different from that observed due to the cohesive stiffening effect of the continuous base.

5.3 Subsurface Instrumentation Layout/Results

Layout

Subsurface instruments of various types were employed to measure the temperature, stress and deformation history of the block during the test program. All instruments with the exception of two extensometers were installed in vertical boreholes drilled from the surface of the block. Two of the extensometer holes were drilled at 30° to the vertical. The borehole instrumentation included: (see Figure 12)

- 80 type K thermocouples
- 2 hollow inclusion triaxial cells (CSIRO)
- 1 three component borehole deformation gage (USBM)
- 2 vibrating wire borehole stressmeters (IRAD)
- 4 four point rod extensometers (MPBX)

Thermal Expansion

The four (two vertical, two inclined) multiple position borehole extensometers (MPBX) were installed primarily to measure thermal expansion in the vertical direction and Poisson expansion in this direction caused by horizontal loading.

During loading at ambient temperature, the anchors generally moved up (toward the surface, positive displacement) or down relative to the deepest anchor (negative displacement) in response to induced lateral stress. The subsurface anchors usually showed negative displacement with increased stress, while the collars were pushed slightly up at the free surface suggesting the existence of a neutral horizon where no vertical displacement took place.

Thermal displacements measured by the MPBXs during confined and unconfined heating of the block were much larger than those induced during load cycles by the Poisson effect. The observed thermal displacements typically increased smoothly during heating, with the anchors moving toward the free surface. When the applied stress was cycled at temperature the anchor movements were usually negative (downward) and exhibited highly hysteritic behavior. The observed thermal displacements were caused by the combined action of confining stress and heating. When either condition was altered (e.g., heater turn-off or cycling of load at temperature) the displacements were reduced. They concluded that interpretation of anchor movement in response to simultaneously changing stress and temperature is a complex three-dimensional problem.

The expansion data derived from the vertical rod extensometers showed the least scatter and most constant results.

An attempt was made to calculate the horizontal (confined) expansion within the block by resolving the rod displacements of the 60° extensometers along the vertical and horizontal planes. The results suggest expansion coefficients approximately 1/3 to 1/2 of those measured in the vertical direction, as well as little or no apparent effect from confinement. Since the extensometers are oriented perpendicular to the termal gradient, the derived results are somewhat questionable.

Thermal Conductivity

Values for thermal conductivity were calculated knowing the heat flux and temperature gradients across the block as monitored from the thermocouples.

Deformations

The two slant borehole extensometers E3 and E4 plunged at 60° in the N-S direction perpendicular to foliation. Deformations recorded by the vertical extensometers E1 and E2 for ambient temperature loading were averaged and the result used to reduce the lateral displacements of the block to the N-S direction from the slant extensometers. The calculated moduli were generally larger than those reduced from surface measurements over similar gage lengths. This was due to the loss of sensitivity to displacement, arising out of the 60° angle between the axis of measurement and the loading direction.

"Poisson" Ratios

"Poisson" ratios of vertical to lateral strain were calculated for the ambient temperature load cycles. To obtain these ratios, average values from the boundary crack analysis (Whittemore pins) were used to estimate lateral strains. During stress cycles at the elevated temperature, the vertical displacements were generally smaller than the accuracy of the extensometers, so strain ratios were not tabulated.

Stress Change Measurements

Two 38mm diameter hollow inclusion CSIRO triaxial stress cells were installed prior to drilling of the flatjack slots. Although the gages appeared to function properly for a time, during or prior to the first ambient biaxial cycle to 6.9 MPa, they apparently became decoupled from the rock.

The 38mm diameter three component USBM borehole deformation gage performed adequately during all test phases, except for an electrical wiring problem that developed in the last unconfined heating cycle. During ambient temperature equal biaxial loading the apparent horizontal N-S and E-W stresses had a ratio 4:1. This result which was, of course, incompatible with the known equal applied stresses, may have been caused by locating the gage in a borehole paralleling a hard quartz lense.

The gage sensitivity factor for the two IRAD vibrating wire stressmeters installed in the block seemed to be heavily dependent on the level of stress induced in the gage by initial installation and the alignment of the installed gage in the borehole. In service, the stressmeters responded somewhat differently than they did during laboratory calibration. Although special attention was given to alignment, the N-S gage (perpendicular to foliation) was roughly 300% sensitive and the E-W gage only 20% sensitive to applied stress.

5.4 Heater Configuration/Test Plan Recommendations

The following recommendations were made by the authors of the Terra Tek report regarding the "heater" aspects of block tests.

- Greater available floor area could be advantageously used for future heated block tests. The greater areal extent of insulated surface would reduce the amount of heat loss to the room. Air and rock temperature measurements indicated that considerable heat loss occurred beyond the insulate surface.

- The heater plane should be extended another two heaters in each direction to insure planes of zero heat flux at the block sides (perpendicular to the heater plane).

- Additional thermocouples should be placed at greater distances from the block to verify any heat loss to the room.

- The use of guard heaters around the block, although expensive, would help to simplify the analysis of conductivity and expansion as a function of temperature. The guard heater array, which could be designed by modelling, would enable one to raise the block to a more or less constant temperature. The central heater array could then be energized at lower power levels to establish a gradient which would allow calculation of the conductivity.

5.5 Summary of Properties and Measurement Technique

An attempt has been made by the present authors to summarize the type of property information derived and the means used to collect the needed information. This has been done in Table III.

Table III. Summary Of the Type and Location Of the Measurements Used to Derive Thermal and Mechanical Properties

Property	Surface	Subsurface	Combined	Not Determined
Thermal Conductivity		T		
Thermal Expansion				
H	HSI,W			
V		MPBX		
Deformation Modulus				
H	HSI,W			
V		MPBX		
Joint Normal Stiffness	W			
Joint Shear Stiffness	W			
Poisson's Ratio				
H	W			
V			W,MPBX	
Stress Change		Data Not Usable		

T = Thermocouples
HSI = Horizontal Strain Indicators
W = Whittemore Pins
MPBX = Multiple Position Borehole Extensometer
H = Horizontal Direction
V = Vertical Direction

As can be seen, much of the usable information has come from surface or near-surface mounted devices. It would appear that considerable attention should be focused on techniques/procedures for gathering property data at greater depths in a heated block than is now possible.

6. SUMMARY

A great deal of information has been collected at the CSM/ONWI Experimental Room regarding

- design of blast rounds in hard rock to minimize damage
- characterization techniques for the rock surrounding an excavation
- permeability measurement/interpretation in unsaturated rocks
- permeability distributions and scale dependence
- the conduct of heated block tests in jointed rock

In addition to techniques/procedures which are non-site specific, a large site specific data base has been amassed. This can be used in modelling studies as well as in future experiment planning.

The highly characterized nature of the site provides a special opportunity for other researchers to evaluate new techniques/ procedures. The site is currently being maintained and potential users are encouraged to contact ONWI/CSM for making arrangements.

Acknowledgements

The authors would like to express their appreciation to the U.S. Department of Energy for sponsorship of the research work. Mr. Richard A. Robinson of ONWI has provided ideas, support and encouragement from program initiation to the present. The CSM graduate students involved in the site characterization work were

Wadood El Rabaa - Borehole Modulus (CSM Cell)
Gideon Chitombo - Crosshole Ultrasonic
Parviz Montzaer - Fracture Permeability and Borehole Logging

The Terra Tek personnel responsible for the Heated Block Test were Howard Pratt. Michael Voegele, Ernest Hardin, Dick Lingle, Mark Board and Nick Barton.

Disclaimer

The views expressed are those of the authors and do not necessarily represent those of the Office of Nuclear Waste Isolation or of the Department of Energy.

APPENDIX A

The titles and status of reports describing the activities at the CSM/ONWI Experimental Room are as follows:

CSM Reports

Report No.	Status	Status
1	CSM/ONWI Hard Rock Test Facility at the CSM Experimental Mine, Idaho Springs, Colorado	Final * Draft
2	Geological and Structural Setting of the CSM/ ONWI Test Site, CSM Experimental Mine, Idaho Springs, Colorado	Final * Draft
3	Hard Rock Excavation at the CSM/ONWI Test Site Using Swedish Blast Design Techniques	Final * Draft
4	Hard Rock Excavation at the CSM/ONWI Test Site Using Crater Theory and Current U.S. Smooth Wall Blasting Practice	Draft
5	An Investigation of Fracture Permeabiltiy Around the CSM/ONWI Experimental Room	Final * Draft
6	Results of Core Logging and Fracture Mapping in the CSM/ONWI Room	Draft
7	An Investigation of Blast Damage Around the CSM/ONWI Room Using a Transient Air Injection Tecnique and a Borehole Petroscope	Draft
8	Blast Damage Assessment Around the CSM/ONWI Room Using Ultrasonic Cross-Hole Measurements	Draft
9	Blast Damage Assessment Around the CSM/ONWI Room Using the CSM Cell	Draft
10	Stress Field In the Vicinity of the CSM/ONWI Room	Draft
11	Rock Stress Measurements by Sleeve Fracturing	Draft
12	Excavation Response	Draft
13	Geometric Contrinuity and Physical Characteristics of the Discontinuities at the CSM/ONW	In Preparation
14	Permeability Measurement and Interpretation For Unsaturated Rock Masses	In Preparation

* Approved For ONWI Release.

References

1. Chitombo, G.P.F., Blast Damage Assessment Using Ultrasonic Cross-Hole Measurements, M.S. Thesis T-2658, Colorado School of Mines, 1982.

2. El Rabaa, A.W.M.S., Measurements and Modeling of Rock Mass Response to Underground Excavation, M.S. Thesis T-2470, Colorado School of Mines, 1981.

3. Montazer, P., Permeability of the Unsaturated Fractured Rocks Near An Underground Opening, Ph.D. Thesis T-2540, Colorado School of Mines, 1982.

4. Hardin, E., Barton, N., Lingle, D., Board, M., and M. Voegele, "A Heated Flatjack Test Series to Measure the Thermomechancal and Transport Properties of In Situ Rock Masses ('Heated Block Test'), Office of Nuclear Waste Isolation Report (ONWI-260), 1981.

5. Holmberg, Roger, "Hard Rock Excavation at the CSM/ONWI Test Site Using Swedish Blast Design Techniques", ONWI Report 140-3, 1981.

Session 2

HYDROLOGICAL INVESTIGATIONS

Chairman - Président

R. BECK
(Switzerland)

Séance 2

RECHERCHES DANS LE DOMAINE HYDROLOGIQUE

HYDROGEOLOGICAL CHARACTERIZATION OF THE STRIPA SITE

J.E. Gale,* P.A. Witherspoon+#, C.R. Wilson+ and A. Rouleau*, *Memorial University of Newfoundland, St. John's; +Lawrence Berkeley Laboratory, and #University of California; Berkeley, California USA

ABSTRACT

A comprehensive program of fracture hydrology investigations has been carried out in a granitic rock mass at the Stripa mine in Sweden. Sources of data included a limited number of outcrops, surface and subsurface boreholes, and fracture maps of rooms and drifts at a depth of about 340 m. The research program consisted of: (1) the collection of fracture geometry and borehole injection test data to determine directional permeabilities, (2) a macropermeability experiment to determine the bulk rock mass permeability, (3) groundwater sampling for investigations of geochemistry and isotope hydrology, (4) pump testing of surface wells, and (5) tracer tests to determine effective porosity. This report summarizes results from all but the last item, the tracer work, which had not been carried out when the field work ended in 1981.

Analysis of a large array of fracture statistics has revealed the existence of four fracture sets. Distributions of spacings and trace lengths were determined for each set. The trace length data are best fitted by a negative exponential distribution while the fracture spacing data are best fitted by a lognormal distribution. A comparison of the borehole injection tests with the results from the macropermeability experiment provides a unique opportunity to evaluate two different methods of measuring permeability in fractured rocks. The average permeability of 1×10^{-13} cm^2 from the macropermeability experiment compares quite favorably with the average value of 8.9×10^{-13} cm^2 determined from borehole testing.

The geochemistry and isotope hydrology data support the concept that the waters found in the granite rock mass at Stripa, especially at the deepest levels (811-838 m), are many thousands of years old. They also support the conclusion that the shallow and deep groundwaters are isolated. Investigations of this kind provide independent approaches to evaluating the degree of isolation in a groundwater system.

The hydrogeological investigations at Stripa have demonstrated the value of coordinated surface and underground testing in site characterization. Proper evaluation of rock mass properties will require detailed studies of the influence of discontinuities, and will involve adequate sampling of their geometric and hydraulic properties from both surface-based and at-depth tests. The work at Stripa has shown that surface and underground testing provide complementary information, and that the complete suite of needed data cannot be obtained from one approach alone.

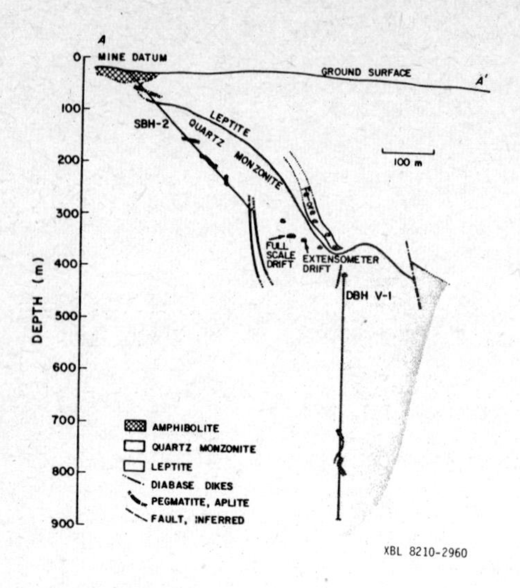

Fig. 1. Vertical cross-section bearing N89°E [from Wollenberg, et al, 4]. See Figure 2 for location.

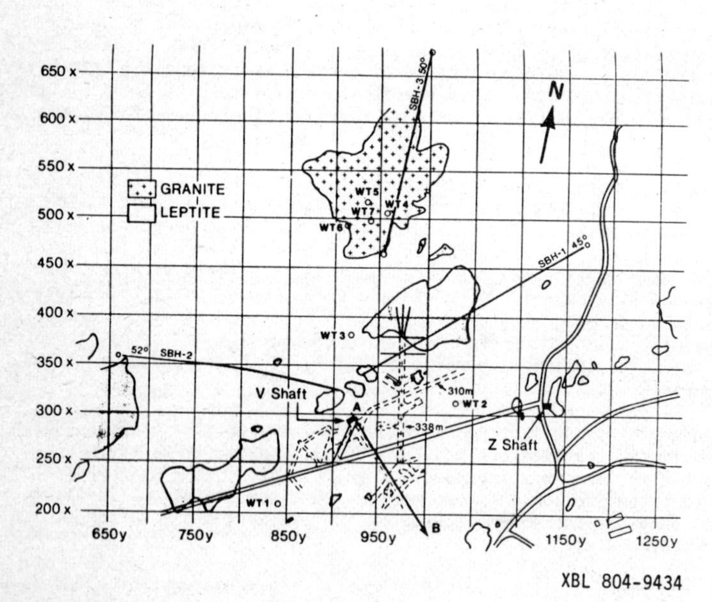

Fig. 2. General geology and hydrology borehole locations. Squares are 50 m on a side.

INTRODUCTION

Increasing interest has been developing in a number of countries in the past few years on the use of granitic rocks for nuclear waste repositories. Thus, when the Stripa mine became available in 1977 [1] providing immediate underground access to a large mass of granite, it was obvious that important problems in waste management could be addressed with a minimum of delay. One of these problems was to understand the factors that control the rate at which radionuclides in groundwater can migrate through rock systems.

The discontinuities are the key factor in addressing the problem of migration through crystalline rock. Flow through the intact rock matrix will be so low that significant movement can only take place through the fracture system. Hence fractures represent the primary flow paths along which radionuclides may migrate from the repository to the biosphere. Therefore to characterize the hydrogeology of a granitic rock mass, one must understand the flow properties of a complex network of fractures.

To investigate the factors that control seepage in a granitic rock mass, we adopted a comprehensive program of fracture hydrology investigations [2]. This program consisted of: (1) the collection of fracture geometry and borehole injection test data to determine directional permeabilities, (2) a macropermeability experiment to determine the bulk rock mass permeability, (3) groundwater sampling for investigations of geochemistry and isotope hydrology, (4) pump testing of surface wells, and (5) tracer tests to determine effective porosity. The report that follows summarizes results from all but the last item, the tracer work, which had not been carried out when the field work ended in 1981.

GEOLOGICAL AND HYDROGEOLOGICAL SETTING

The Stripa mine site is located in south central Sweden about 150 km west-northwest of Stockholm. The bedrock geology is typical of highly folded and deformed shield terrains. The regional geology is characterized by a northeast-southwest trending suite of folded metamorphic rocks that have been intruded by a series of granitic rocks. The local bedrock structure around Stripa is dominated by a northeast-southwest trending syncline. Additional smaller synclines, one of which contains the Stripa ore zone (Figure 1), trend both parallel and perpendicular to the major southwest trending syncline and add to the overall structural complexity of the region. Superimposed on the regional fold pattern is a series of fracture zones and lineaments with at least one major fracture zone following the trend of the major synclinal feature [3].

The experimental rooms for much of the research program were excavated at a depth of 338 m below surface under the north limb of the syncline (Figure 1), in a small body of granite (quartz monzonite) adjacent to the metasedimentary - metavolcanic sequence in which the mined-out ore body was located. The general geology of the test site area and the general fracture system are described by Olkiewicz et al. [3]. The petrology of the granite body is discussed by Wollenberg et al. [4].

Sources of data for the fracture geometry investigation included a limited number of surface outcrops (Figure 2), three oriented surface boreholes, fifteen subsurface hydrology boreholes, and a large number of boreholes drilled for the thermomechanical experiments [5, 6]. Fracture maps of walls and floors of experimental rooms, especially the maps of the drift [7] for the macropermeability experiment, provided an additional important source of data.

The surface boreholes used in the fracture hydrology work consisted of three long (315-385 m) inclined boreholes shown as SBH-1, SBH-2, and SBH-3 on Figure 2. These three boreholes were oriented to optimize their intersection with the major fracture sets [2]. The subsurface hydrology boreholes consisted of 15 diamond coreholes 30-40 m in length. They were drilled at the north end of the underground test rooms within the quadrant shown in Figure 2 bounded by mine coordinates 350-400 in the x direction and 950-1000 in the y direction. The last 33 m of this main drift was used for the macropermeability experiment [8, 9]. All boreholes were 76 mm in diameter, and the drillcores were oriented and reconstructed to enable determination of true fracture strikes and dips.

An extensive series of measurements and observations were made in the boreholes. These included: (1) in situ pore water pressure measurements during drilling, (2) measurement of the variation in water levels in boreholes open at the surface, (3) water outflow measurements from instrument and heater boreholes in conjunction with the thermal experiments, (4) measurement of water pressures and flow rates during both injection and withdrawal packer tests, and (5) measurement of water levels during pumping out tests in several surface water wells. The raw data from these observations and measurements as well as the methods used to collect and process both the fracture and hydrology data have been described [10]. They serve as the basis for this investigation of the fracture and hydrogeological characteristics of the Stripa site.

The most significant perturbation of the hydrologic regime at Stripa has been that produced by excavations made during the mining operations. Mining started as an open pit operation some 400 years ago and continued as an underground operation for about the last 40 years. In the initial phases of the Stripa program it was proposed that the mine was acting as a major drain, decreasing the pore pressures in the surrounding rock mass. Thus, an attempt was made to measure in situ pore pressures during the drilling of the three surface wells SBH-1, SBH-2 and SBH-3.

The in situ pore pressures for SBH-1 and SBH-3 are plotted in Figure 3. The deviation of the fluid pressures from hydrostatic conditions in SBH-1 indicates a strong downward gradient starting about 100 m below ground surface. During the drilling of SBH-2 no measurements were made below the 100 m level. However the data from SBH-3 show downward gradients starting at the surface. These hydraulic gradients confirmed the expected drainage effect of the old mine workings. Moreover, the measurements made by the Swedish Geological Survey in DBH V-1 (Figure 1), which is a flowing borehole at the 410 m level, indicate upward acting gradients which suggest that groundwaters from a deep flow system are discharging upward into the mine.

CHARACTERISTICS OF FRACTURE SYSTEM

To determine the characteristics of the fracture system, data have been obtained from: (1) surface and subsurface hydrology boreholes [10], (2) boreholes drilled for the thermomechanical experiments [3, 5, 6], and (3) maps of the walls and floor of the ventilation drift where the macropermeability experiment was conducted [7]. The drill cores from both surface and subsurface boreholes were

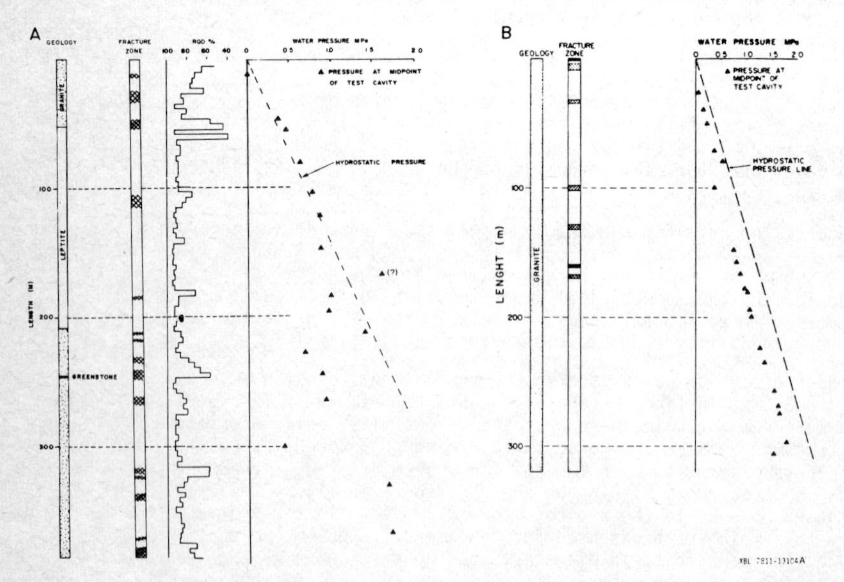

Fig. 3. Summary of geology, fracture data, and in situ fluid pressures in boreholes SBH-1 (A) and SBH-3 (B).

oriented and hence true orientations of the fractures intersecting the drill core were calculated. These different data sets have been combined to define the orientation of the different fracture sets within the test area and their trace length and spacing distributions.

Fracture Orientations

Rouleau and Gale [11] have made a detailed analysis of the fracture measurements from which the geometry of the fracture network has been deduced. Approximately 10,000 individual fracture measurements were used in this analysis. Figure 4 summarizes the fracture orientation data in the form of lower hemisphere, equal area, contour plots of poles to fracture planes. Individual plots are given for data from the full scale drift, time scale drift, the R and HG holes, the walls and floor of the ventilation drift, and the lower part as well as the total length of the three SBH boreholes. It should be noted that a considerable bias is introduced into these data by the orientation of the boreholes or the surfaces from which the fracture data were collected.

The contour diagram in Figure 4 for the time scale drift was constructed using only vertical borehole data [5] while the data used to construct the

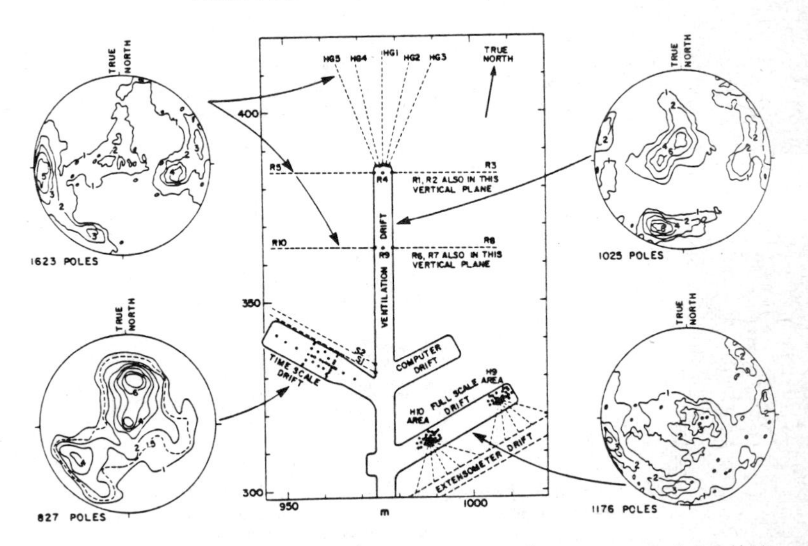

Fig. 4. Contoured stereonets of poles to joint planes measured in different areas of test excavation. The contoured values are in percent of points per 1% surface area [from Rouleau and Gale, 11].

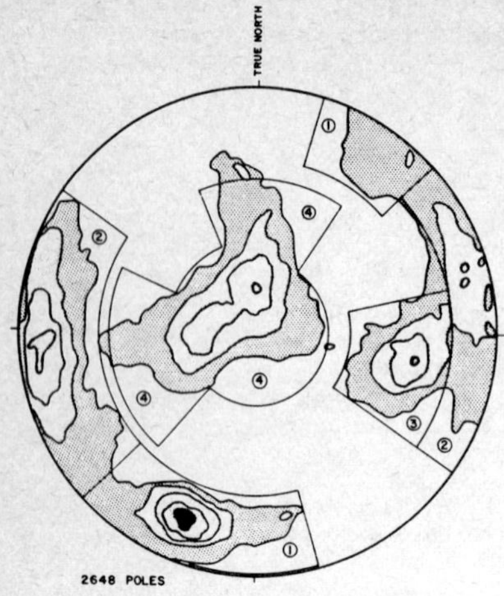

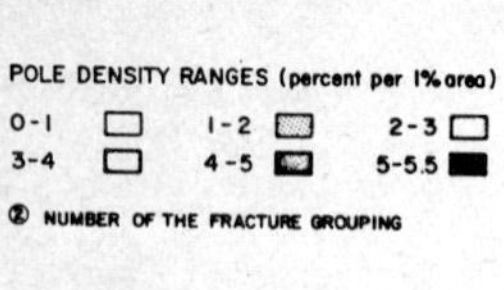

Fig. 5. Lower hemisphere plot of poles to fracture planes for walls of ventilation drift and HG and R drill cores [after Rouleau and Gale, 11].

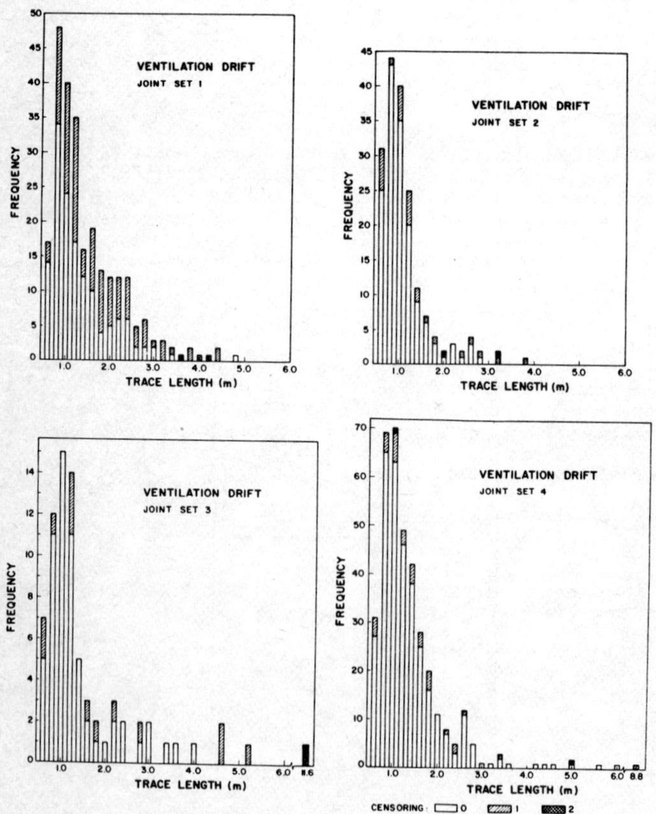

Fig. 6. Frequency histograms of the fracture trace lengths measured on the walls of the ventilation drift [after Rouleau and Gale, 11].

diagram for the full scale drift were obtained from both vertical and horizontal boreholes. In areas of complex fracturing and folding, such as the Stripa site, more scatter is introduced into the data as the volume of rock increases. This is evident when we compare the two contour plots for the SBH boreholes. When data from the entire length of the SBH boreholes are included, the fracture sets are very poorly defined or missing, with considerable scatter evident in the result. However, using data from only the bottom portions of the three SBH boreholes (i.e. from depths greater than 175 m) results in a significant reduction in data scatter and considerable improvement in the definition of the fracture sets.

Figure 5 is a contour plot of fracture data from both R and HG holes as well as the data from the walls and floor of the ventilation drift (see Figure 4). Given the variation in sampling orientation provided by the boreholes and the mapped surfaces, this plot can be assumed to give a relatively unbiased sample of the fracture system in the immediate area of the ventilation drift. Four fracture sets have been defined and their linearized boundaries superimposed on the contour diagram (Figure 5). Visual inspection shows that the pole clusters for these four sets correspond fairly closely with the clusters defined on the orientation diagram for the bottom part of the SBH holes (Figure 4).

Fracture Trace Lengths

Having obtained the fracture orientations, the next step was to investigate trace lengths. During the mapping of the ventilation drift, the length of each fracture trace was measured [7]. Based on its known orientation, each fracture was assigned to one of the fracture sets identified in Figure 5. Then for each fracture set, the trace lengths have been plotted as frequency histograms (Figure 6). In addition to showing the distribution of fracture lengths, Figure 6 also indicates the degree of censoring, i.e. 0 when both ends of the fracture trace are visible on the mapped surface, 1 when only one end is visible, and 2 when neither end is visible. A summary of the statistics computed for the trace length distributions is given in Rouleau and Gale [11].

The censoring and other sampling biases such as truncation and size bias [12, 13] make the statistical analysis of trace length distributions relatively difficult. A maximum likelihood approach to estimate the parameter of an exponential distribution accounting for censoring [14] has been applied to our data. The probability density function of an exponential distribution is given by:

$$f(x) = \lambda e^{-\lambda x} \tag{1}$$

The mean of this distribution is $1/\lambda$. Although no test has been carried out for the goodness-of-fit of this distribution to the data, the results suggest that there is a significant difference in trace lengths from one fracture set to another. For example, the computed estimates of the mean values of trace lengths for the four different sets range from 1.3 to 2.7 m. The complete statistics for trace length data are given in Table 1.

Table 1. Trace length statistics by degree of censoring for the four fracture sets.

	SET 1			SET 2			SET 3			SET 4		
CENSORING[1]	0	1	2	0	1	2	0	1	2	0	1	2
N	140	109	2	150	26	2	61	13	1	327	35	2
SUM	181.55	183.25	7.89	158.12	36.38	5.04	82.40	28.79	11.51	435.42	59.14	5.96
MAX, m	4.78	4.44	4.28	2.93	3.85	3.11	3.95	5.22	11.51	6.04	8.88	5.02
MEAN, m	1.30	1.68	3.95	1.05	1.40	2.52	1.35	2.22	11.51	1.33	1.69	2.98
STD. DEV., m	0.68	0.86	----	0.45	0.88	----	0.77	1.71	-----	0.73	1.53	----
SKEWNESS, m	1.76	1.15	----	1.78	1.42	----	1.76	0.98	-----	2.58	3.53	----
KURTOSIS, m	4.48	1.00	----	3.52	1.34	----	2.60	-0.48	-----	10.23	14.77	----
$\hat{\theta}$		0.376			0.752			0.497			0.653	
$1/\hat{\theta}$[2]		2.660			1.330			2.012			1.531	
$V[\hat{\theta}]$[3]		0.0010			0.0038			0.0041			0.0013	

Note: 1. Degree of censoring is 0 when both ends of the fracture trace are visible, 1 when only one end is visible, and 2 when neither end is visible.
2. Theoretical measure of the mean of the exponential distribution [11].
3. Variance.

Fracture Spacings

The spatial locations of the fractures intersecting the drill cores have been used to compute the spacing between every pair of consecutive fractures of the same set. We define the spacing here as the distance between consecutive intersections of two fractures of the same set with a sampling line (i.e. a borehole axis), multiplied by the cosine of the angle made by the sampling line and the pole of the average plane of the fracture set. Figure 7 shows the frequency histograms of spacings for every fracture set that has been defined for the rock mass surrounding the ventilation drift [11]. The spacing data from all the oriented HG and R drillcores were combined to construct the histograms of Figure 7. A summary of the statistics computed for these spacing distributions is given in Table 2.

Table 2. Summary statistics of the distributions of spacing (SPAC) and logarithm of spacing (LSPAC) for each fracture set for oriented HG and R holes.

	SET 1		SET 2		SET 3		SET 4	
	SPAC	LSPAC	SPAC	LSPAC	SPAC	LSPAC	SPAC	LSPAC
N	209	208	619	614	177	177	319	318
MAX, m	7.08	1.96	4.42	1.49	8.59	2.15	9.29	2.23
MIN, m	0.0	-5.51	0.0	-6.21	0.003	-5.82	0.0	-4.85
MEAN, m	0.93	-0.88	0.36	-1.74	0.79	-1.08	0.51	-1.50
STD.DEV., m	1.21	1.40	0.54	1.23	1.25	1.36	0.92	1.26
SKEWNESS, m	2.42	-0.34	3.58	-0.11	3.51	-0.17	5.21	0.15
KURTOSIS, m	7.03	-0.22	16.50	0.23	15.00	0.25	37.13	0.04
WEIBULL, m								
Shape (λ)	.830		.856		.797		.798	
Scale (σ)	.769		.317		.625		.408	
MODEL	D-STATISTICS[1] AND [P(>D)][2]							
Exponential	.126 [.002]		.126 [<.001]		.182 [<.001]		.200 [<.001]	
Normal		.043 [>.15]		.019 [>.15]		.048 [>.15]		.055 [>.15]
Weibull	.043 [>.15]		.066 [<.01]		.075 [>.15]		.095 [<.01]	

Note: 1. Measure of goodness of fit for assumed model according to Kolmogorov-Smirnov D statistics [11].
2. Probability that value of D is greater than observed value if the empirical distribution was exactly following the theoretical model.

Based on the shape of the histograms in Figure 7, these empirical distributions have been compared to various theoretical models that are bounded by zero to the left and skewed to the right. The distributions were tested against exponential, lognormal and Weibull models. The analysis and goodness-of-fit tests presented in Rouleau and Gale [13] indicate that the exponential distribution does not fit the data at all. For two of the four fracture sets the Weibull model passes the test at a level of significance larger than 0.05. However, the lognormal distribution fits the data very well at a level of significance larger than 0.15.

In order to assess variability in the fracture spacing around the ventilation drift, a one-way analysis of variance was made to test the hypothesis that, for each fracture set, all the populations of spacings sampled from the different drill cores have the same mean. This analysis indicated that, except for Set 3, the fracture sets definitely do not have the same population mean [13]. This suggests that each individual borehole samples a volume of rock that is smaller than the scale of variability in the density of fracturing for the whole rock mass.

A similar analysis was carried out using the following borehole groups: (1) HG boreholes, (2) R1, R2, R3 and R5, and (3) R6, R7, R8, and R10. Results of this analysis did not indicate significant differences between the population means, except for Set 4. These results suggest that each of these groups sampled a volume of rock that is larger than the scale of variability in the density of fracturing for the whole rock mass. Thus, combining the spacing values from the HG and R boreholes for all the fracture sets should yield spacing distributions that can be considered representative of the rock mass in any portion of the ventilation drift.

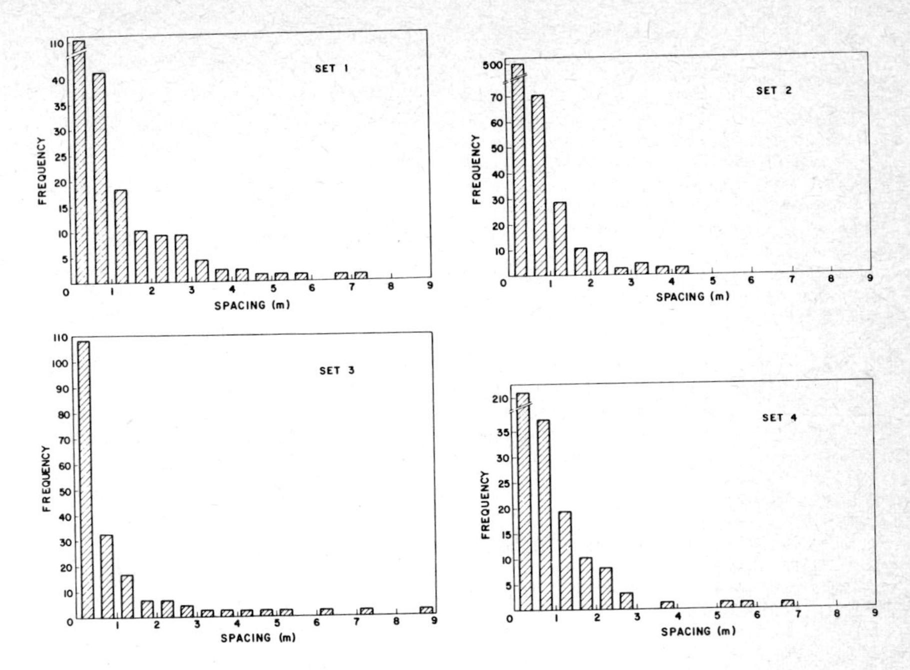

Fig. 7. Frequency histograms for the spacing between consecutive fractures of the same set measured on the oriented cores from HG and R boreholes [from Rouleau and Gale 11].

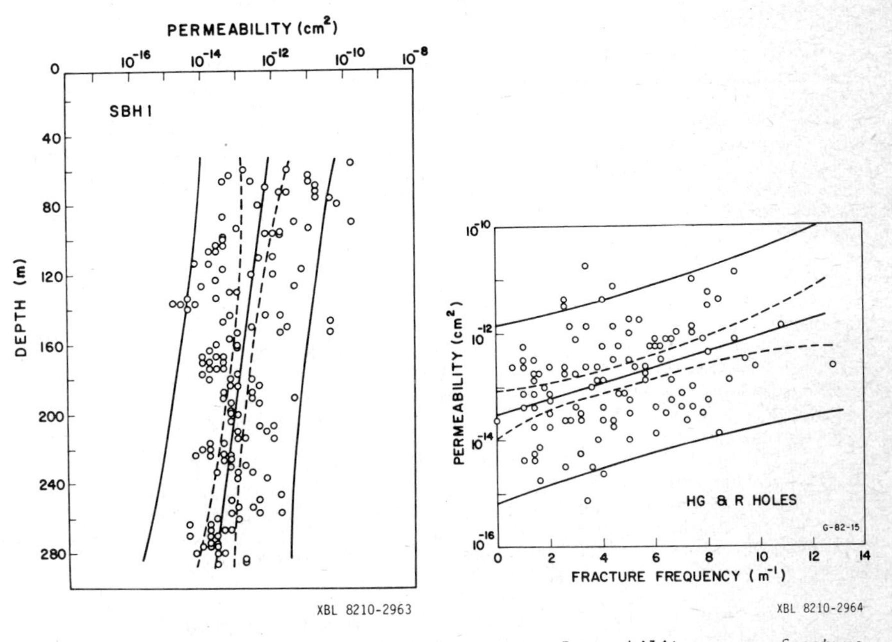

Fig. 8. Permeability versus depth from injection tests in SBH-1 [after Gale, 15].

Fig. 9. Permeability versus fracture frequency for all HG and R boreholes [after Gale, 15].

HYDROGEOLOGICAL RESULTS FROM BOREHOLES

Permeability

Injection tests were made with packed off intervals 2 m long over the total lengths of the three SBH surface boreholes and also over 2m and 4m lengths in the R and HG subsurface boreholes. A detailed discussion of the equipment and testing procedures is given by Gale [10]. A total of about 850 intervals were tested with many of the intervals being subjected to several injection tests at different pressures. Equivalent porous media permeabilities have been computed for each test interval, and the results from the 2m spacings in SBH-1 are shown in Figure 8. While there is considerable scatter in the test results, a least squares fit to the data does indicate that there is a general trend of decreasing permeability with increasing depth [15].

The effect on permeability of the number of fractures intersecting the test interval in the HG and R boreholes is shown on Figure 9. This figure shows that with an increase in the number of fractures intersecting a test interval (increase in fracture frequency), there is a general increase in the permeability. However, one would have anticipated that if all fractures were contributing equally to the flow behavior, there should have been a much stronger correlation.

The overall variation in rock mass permeabilities in the walls of the ventilation drift is shown by the histogram in Figure 10. This histogram was compiled using data from 148 individual tests in the HG and R boreholes [15]. The geometric mean is 8.9×10^{-13} cm^2.

Fracture Porosity

One of the fundamental problems in fracture hydrology is to determine the volume of pore space that controls fluid movement. This is called the "effective", or "flow", porosity but there can also be a significant amount of pore space that does not affect fluid movement because of the dead end spaces. The "total" fracture porosity includes all fracture openings whether they contribute to the flow process or not.

Fracture porosity, at least with regard to the crystalline rocks that are being considered for waste isolation, is normally far less than the matrix porosity. Hard crystalline rocks, for example, have matrix porosities of 0.01 to 0.02 [16], whereas fracture porosities are typically less than 10^{-4}. Thus, there are inherent problems in measuring fracture porosities in the field especially when the methods are influenced by the relatively large volume of pore space in the matrix. As a result, there are not many insitu determinations of fracture porosity.

A new procedure is being developed to determine fracture porosity using the wealth of data collected at Stripa. Figure 11 is a probability plot of apertures that were computed using the injection test data from the HG and R boreholes. The data points on the right side (solid circles) represent results from about 100 tests where single fractures were isolated and tested. The second set of points on the left side (open squares) represent results from about 120 tests using a fixed spacing of 2 m. The computed results for this second correlation were based on the assumption of a single fracture within the 2m interval regardless of the actual number.

In both correlations on Figure 11, the aperture results are reasonably close to a straight line, and we interpret this to indicate that the fracture apertures are lognormally distributed. Based on this interpretation, we have computed mean values of 5.8 micrometers for the single fracture data and 8.3 micrometers for the 2m interval tests.

The apertures in Figure 11 were calculated from the injection test data using the standard assumption of smooth parallel walls, but it can be demonstrated that this may introduce an error caused by ignoring the effects of roughness

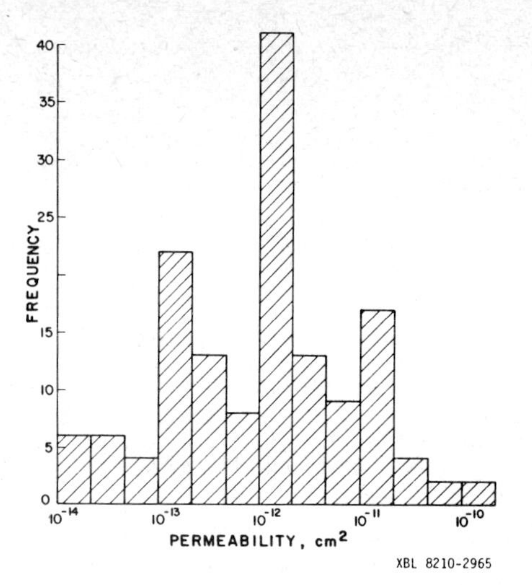

Fig. 10. Permeability distribution computed for the HG and R boreholes [after Gale. 15].

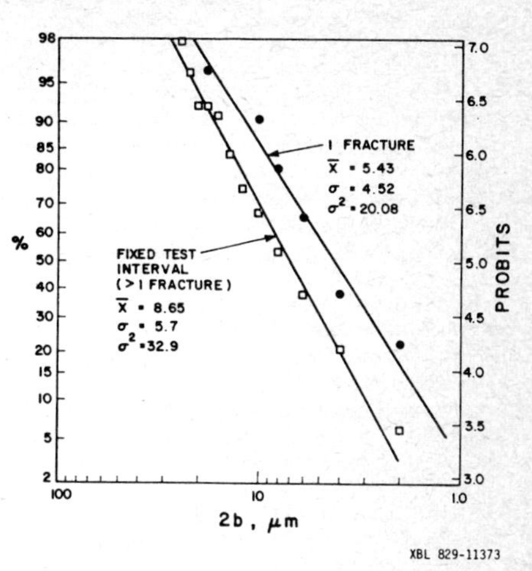

Fig. 11. Distribution of fracture apertures from 2 m and single aperture injection test data collected in the HG and R boreholes [after Gale, 15].

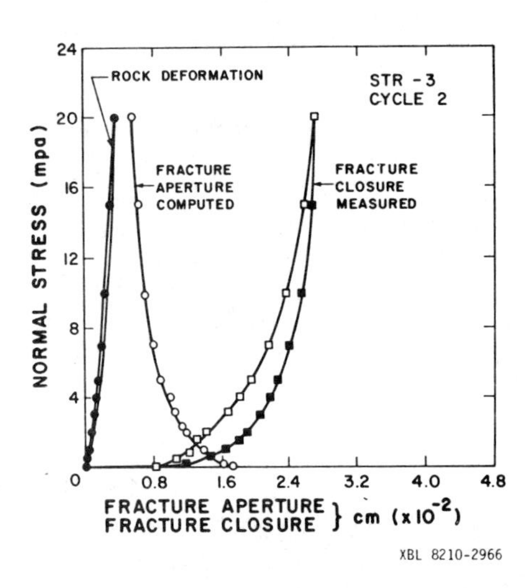

Fig. 12. Results of laboratory testing on natural fracture in a Stripa core sample showing effect of normal stress on fracture behavior.

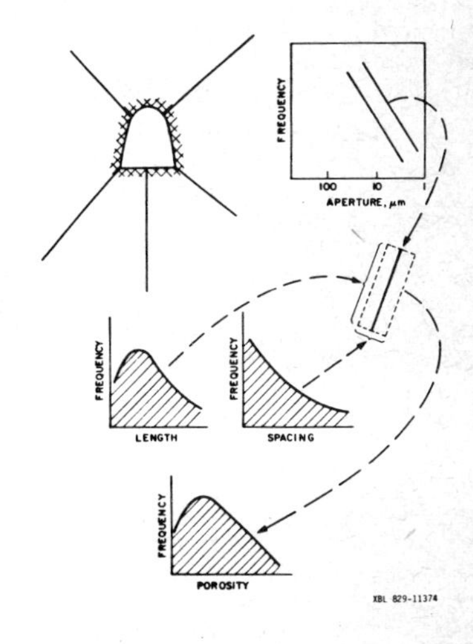

Fig. 13. Graphical representation of method for computing fracture porosity from aperture data and fracture statistics.

[15]. Furthermore, the calculated results depend on the degree of interconnection between the fractures within the test interval and the rest of the network.

Some idea of the actual size of the fracture apertures has been obtained from an analysis of results [15] obtained in laboratory testing of a natural fracture in a core of Stripa granite with a diameter of 0.15m. In the laboratory situation, the single fracture is continuous between the imposed pressure boundaries, and hence there is no problem with the effects of fracture interconnection. Figure 12 shows the relationship between measured fracture closure and normal stress across the single fracture. Apertures were computed from the flow data on the assumption of smooth parallel walls.

The in situ injection tests in the HG and R boreholes were conducted in the rock mass where the estimated in situ stress is in the range of 10-20 MPa. From Figure 12, this corresponds to a parallel-plate fracture aperture of about 80 micrometers, and if the effects of roughness were taken into account, this aperture would be somewhat larger. Thus, once corrections for roughness are made, the field data on Figure 11 and the laboratory data on Figure 12 can be used to develop a fracture distribution model for this granitic rock mass.

The final step in this new procedure is to combine the aperture data with the fracture statistics to determine the effective and total porosities. Figure 13 illustrates the approach that we are currently investigating. The process is as follows: (1) identify the set to which each fracture belongs, (2) from the statistical correlations for that set, randomly select a trace length and spacing for the given fracture, and (3) randomly select an aperture for each fracture from the fracture distribution model. One can then estimate the <u>total</u> porosity using aperture data as interpreted from laboratory results (Fig. 12) and <u>effective</u> porosity using aperture data as computed from borehole tests (Fig. 11). This procedure has been used to determine fracture porosities for six of the 15 boreholes in the ventilation drift [15]. The results indicate that the rock mass has a mean effective fracture porosity of the order of 10^{-5} and a mean total fracture porosity of the order of 10^{-4}. Further work is needed to verify the applicability of this new approach in evaluating fracture porosity.

Pump Testing of Surface Wells

A cluster of four wells 15-30 m apart was drilled into a surface outcrop of granite approximately 100 m north of the underground test facility to determine if an indication of directional permeability could be obtained from pump testing. The wells are shown as WT4, WT5, WT6 and WT7 on Figure 2 and they were located based on the orientations of the dominant fracture sets. The central pump well, WT7, was 100 m deep and the observation wells were each 50 m deep. All wells were vertical. Test results indicated the local hydrologic system was dominated by a fracture zone interconnecting the pumped well with one of the observation wells. The other two observation wells did not appear to be hydraulically connected to the pump well. The dominant effects of this single fracture zone did not permit the test results to be interpreted in terms of directional permeability and suggest that the wells may not have intersected a representative number of fractures. Additional information on these tests has been published [17].

GEOCHEMISTRY AND ISOTOPE HYDROLOGY

Geochemistry and isotope hydrology of groundwaters provide an independent approach to the problem of the overall permeability of a rock system. If surface waters moved rapidly to the experimental level (338 m), shallow and deep waters should be similar in chemistry and age. On the other hand, if the deep waters entered the groundwater system many thousands of years before the shallow waters or at a considerable distance from the present site, to which they percolated slowly, there should be significant differences between waters at different depths.

A comprehensive program of geochemical investigations of the Stripa ground waters has been carried out by Fritz et al. [18, 19]. Water samples were collected from the surface, shallow private wells, and boreholes drilled in the heater drifts at the 338 m level. In addition, samples were collected from a deep borehole drilled by the Swedish Geological Survey from 410 m (the deepest operating level in the mine) to about 840 m below the surface.

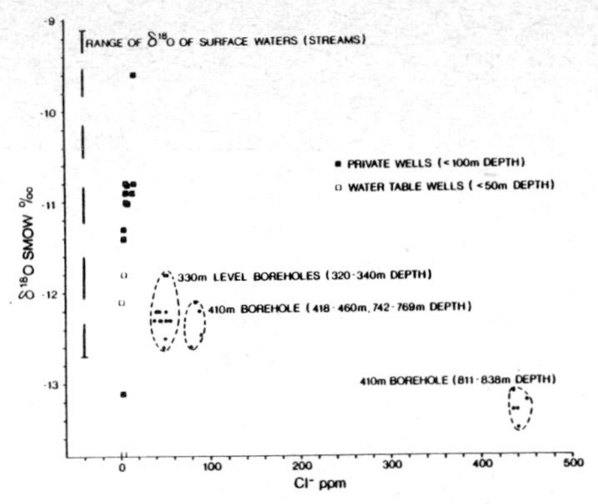

Fig. 14. Comparison of chloride concentrations with $\delta^{18}O$ values from geochemical investigations shows there are distinct differences between waters at different depths. Oxygen isotope values are referred to standard mean ocean water (SMOW).

The high chloride concentrations and relatively low deuterium and oxygen-18 contents of the deep mine waters, in comparison to near-surface waters, indicate that these waters followed different flow paths and/or had a different geochemical history. The low deuterium and oxygen-18 values also indicate that when the deep waters were originally at the surface (that is, before they seeped into the groundwater system), their temperatures were considerably cooler than that of present day conditions. The high chloride contents suggest a much different geochemical environment than presently exists at Stripa. These conclusions are substantiated by comparing $\delta^{18}O$ with the chloride concentrations, as shown in Fig. 14. It is apparent that the deep ground waters, especially those at 811 to 838 m, are distinctly different from the shallow groundwaters. This is interpreted as an indication that the shallow and deep groundwater systems are hydrologically isolated from each other.

Isotopic dating of the various groundwaters was also carried out [18, 19]. In contrast to the surface waters, where appreciable amounts of tritium were observed, the deep groundwaters from the quartz monzonite are essentially devoid of tritium, which indicates that they are at least 30 to 40 years old. Waters from the deep levels are also very low in dissolved inorganic carbon; 2000 to 3000 liters of water were needed for ^{14}C analysis. On the basis of this method, the age of the waters at the 330 m level, and probably also from the 410 m borehole, exceeds 20,000 years.

Three different approaches to dating based on the uranium decay series were also investigated; these involved: (i) uranium activity ratios, (ii) helium contents, and (iii) radium-radon relations [18, 19]. Although the $^{234}U/^{238}U$ method is still under development and is subject to some uncertainties, ages exceeding 100,000 years have been inferred from these data. Similar ages have been determined from ^{4}He concentrations. A method proposed by Marine [20], relating ^{4}He to its parent ^{238}U, yields ages ranging from tens of thousands to hundreds of thousands of years. The results of the radium-radon method indicate ages for the groundwaters ranging from 8,000 to 35,000 years.

Both the ^{14}C and the uranium decay series are dating tools which consider dissolved constituents in the water rather than the water itself. Thus, geochemical reactions can affect isotopic distributions and water ages. However, despite this the data support the concept that the waters found in the quartz monzonite rock mass at Stripa, especially at the deepest levels (811 to 838 m), are indeed many thousands of years old. They also support the inference from the geochemical differences cited above that the shallow and deep groundwaters are isolated. It is apparent that geochemical and isotope hydrology investigations provide independent approaches to evaluating the degree of isolation in a groundwater system.

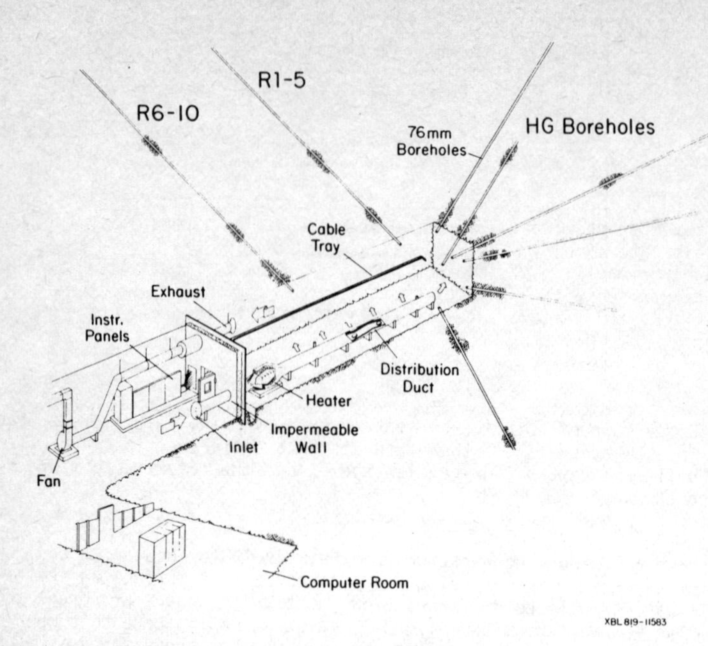

Fig. 15. Schematic drawing of the macropermeability experiment in the ventilation drift showing installed equipment and boreholes that were packed off to obtain pressure measurements.

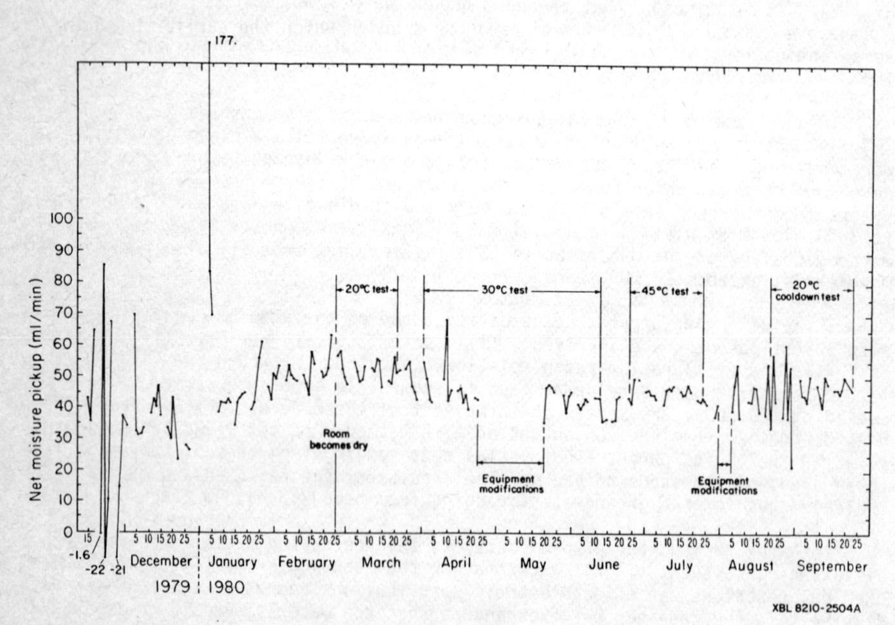

Fig. 16. Net moisture pickup by ventilation system for period 15 November 1979 to 30 September 1980. Lines join data taken on consecutive days.

THE MACROPERMEABILITY EXPERIMENT

The macropermeability experiment was an attempt to measure the permeability of a large volume of fractured rock. The experiment was conducted in a 5 m x 5 m x 33 m room known as the ventilation drift (see Figure 4) at the 338 m level of the Stripa mine. Water flow and pressure were monitored for an 11 month period from November 1979 to September 1980.

The primary objective of the experiment was to improve techniques for permeability characterization of large volumes of low permeability rock. In meeting this objective, the results were expected to: (1) give a measure of the average permeability of some 200,000 cubic meters of rock, (2) provide a basis of comparison with estimates of rock mass permeability from the conventional borehole tests, and (3) evaluate the ventilation technique for monitoring seepage into a large, open drift.

Discussions of the theoretical basis for the experiment and the need for this type of research are presented in earlier papers [2, 8, 21, 22]. An analysis of the steady state, radial component of seepage into the drift has also been carried out [9].

A schematic drawing of the experimental setup is presented in Figure 15. Water inflow was measured as the net moisture pickup of the ventilation system inside a sealed portion of the ventilation drift. The ventilation air was heated to improve its moisture carrying capacity and to improve the accuracy of the measurements. Hydraulic gradients were determined by measuring water pressure in 90 isolated intervals in the 15 HG and R boreholes (Figure 15). Each isolated interval was about 5 m long, a length intended to include sufficient numbers of open fractures (generally 15 to 20) within each zone to provide reasonable assurance that the pressure data would be sufficiently averaged to produce smooth pressure profiles.

Tests were run for three different nominal room temperatures of 20°, 30°, and 45°C, followed by a cooldown test back to 20°C. The 20°C test most closely reproduced the ambient air conditions under which the earlier borehole permeability tests were performed in the R and HG boreholes of the ventilation drift. The two higher temperature tests provided a higher accuracy measurement and permitted investigation of thermal effects on the measured inflow and on the ventilation techniques for monitoring seepage. The first three tests were continued until steady state conditions were essentially reached. Because of time constraints the cooldown test was stopped while still clearly in a transient state.

Net moisture pickup from the ventilation system measurements is shown in Figure 16 for the period 15 November 1979 to 30 September 1980. Net moisture pickup was determined from the difference between the water contents of air streams in the exhaust and inlet ducts. It was interpreted as a measure of the average seepage rate into the sealed portion of the ventilation drift.

Net moisture pickup for the initial 20°C test averaged about 50 ml/min, although the data show considerable scatter due to low measurement accuracy at this temperature. This first test was actually run at an average room temperature of about 18°C, which was as close to the ambient mine air temperature of 15°C as could be attained while still evaporating all incoming moisture. Figure 17 shows a distance-drawdown, semi-log plot of the pressure data from the radial boreholes. On such a plot the normal porous media type flow will appear as a straight line. Data from four of the ten individual boreholes do plot as rather well defined straight lines, with correlation coefficients on the order of 0.9 or better. These are shown by the dashed lines on Figure 17. The weighted average of all data points also plots as a straight line, with a correlation coefficient of 0.98. Assuming that 80% of the observed 50 ml/min net moisture pickup occurred as radial flow (as estimated from a simplified analysis of flow into the entire drift), the average hydraulic conductivity of the monitored rock mass is about 1.0×10^{-10} m/s, which for water at ambient conditions is equivalent to an intrinsic permeability of 1.0×10^{-13} cm^2.

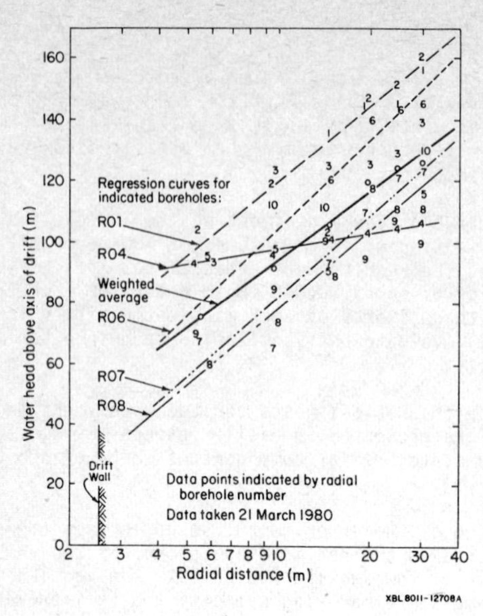

Fig. 17. Distance-drawdown plot for permeability test with nominal room temperature of 20°C.

It may be noted on Figure 17 that the weighted average line and the lines for each borehole indicate a water head significantly higher than the actual head if projected to the wall of the drift. This suggests that a skin of lower permeability rock exists between the drift wall and the location of the closest measurements. If this skin is assumed to be homogeneous and about 2.5 m thick, its hydraulic conductivity would be about 3.5×10^{-11} m/s, or about one-third of the average conductivity of the rock mass. If the skin thickness were less, its conductivity would be even smaller.

At the end of the 20°C test, air temperatures were raised successively to 30°C and 45°C and then reduced back to 20°C. Table 3 summarizes the permeability results obtained from the four tests, and it will be noted that essentially the same results were obtained at all temperatures. It was observed that groundwater pressures tended to increase in the vicinity of the drift with increasing air temperatures. This resulted in slightly smaller gradients which were reflected in the decreased flow rates. The mean hydraulic conductivity for all the tests is 9.8×10^{-11} m/s, which is equivalent to an intrinsic permeability of 9.8×10^{-14} cm^2.

Table 3. Results of permeability tests in the macropermeability drift.

Test	Nominal Temp. °C	Average Flow Rate ml/min	Hydraulic Conductivity m/s
1	20	50	1.0×10^{-10}
2	30	42	9.4×10^{-11}
3	45	43	1.0×10^{-10}
4	20[1]	47	9.8×10^{-11}

Note: 1. Cooldown Temperature

DISCUSSION AND CONCLUSIONS

Given the importance of the fracture system, both in terms of mine stability and groundwater flow, it is obvious that a careful analysis of the fracture system is mandatory if we are to compare the characteristics of one site with those of another. This becomes even more important when we consider the number of sites that are currently being studied in different countries. Based on our work at Stripa, several points can be made that will be of assistance in selecting a repository site.

The data needed to evaluate fractured rock masses are quite considerable, and their collection and preservation are of utmost importance. To answer questions relevant to site assessment, it is clear that a careful and systematic method of field operations must be adopted. Maps of the drift walls should be user oriented. This involves careful photographing of drift surfaces, followed by mapping the different fracture traces on a grid system. Each fracture trace should be assigned a number so that the relevant data, such as strike, dip, trace length, etc., for that particular fracture can be located in the data files. This will make it relatively easy for other workers to use the basic fracture maps and data files to compare fracture characteristics at one site with that at another site.

The drill core is another source of valuable basic data. The boreholes should be oriented to optimize intersection with the principal fracture sets. The cores should be oriented, photographed and carefully mapped so that the true orientations of the fractures intersecting the drillcore can be calculated. This requires the use of improved core orientation devices, triple-tube core barrels, and impression packers in conjunction with a sustained effort of core reconstruction and mapping.

At Stripa user-oriented maps [7] were prepared for the ventilation drift. These maps were used to determine the number of fracture sets present in the rock mass and the distribution of trace lengths for each fracture set. Trace lengths can only be measured at surface exposures or on the walls and floors of subsurface drifts. A sufficient area of exposure is necessary to minimize the censoring problem and to insure that an adequate sample has been obtained.

The oriented HG and R drill cores provided an independent set of data that confirmed the existence of the four different fracture sets. In addition we were able to use the oriented drill cores to determine the distribution of spacings for each fracture set. Analysis [11] of the fracture data from the drift walls suggests that the trace length data are best fitted by a negative exponential distribution while the fracture spacing data from the oriented drill cores are best fitted by a lognormal distribution. This analysis also showed that significant differences exist between the mean values of the trace lengths and spacings for the different fracture sets.

It is obvious that reliable data are required from both surface and subsurface exposures and oriented drill cores if one is to determine the type of statistical distribution that best fits the trace length and spacing data from different sites. It is essential that the appropriate distributions be identified so that their statistical parameters may be used in simulations of fracture networks. This will enable one to assess flow pathways, hydraulic interconnections, directional permeabilities, and fracture porosities. Fracture data will also provide valuable input for the general problem of rockmass stability and questions of thermomechanical behavior. In this context, the new approaches used in collecting and analyzing the fracture data at Stripa, when modified by the experience gained, should provide a statistically meaningful basis for site comparisons.

The fracture hydrology program at Stripa was also designed to explore new techniques for evaluating the fluid flow properties of discontinuous rock masses of low permeability. A fundamental question that is inherent in such evaluations is whether or not the fractured rock can be analyzed using porous media concepts. What measurements should one make to verify that a porous media equivalent exists for a network of discontinuous fractures? To provide new insight into this problem, the field program at Stripa was deliberately designed to include permeability measurements on a wide scale ranging from individual fractures to the very large scale macropermeability experiment. Further work is needed, however, to analyze the fracture statistics in order to shed more light on the question of the porous media equivalence of the fracture network. New methods of investigating this problem are now available [11, 23]. Fracture porosity is another important component of this problem. Porosity values can be computed using the techniques discussed above, but the field measurement of effective porosity will require appropriate tracer studies.

The testing equipment that was developed for the fracture hydrology measurements in boreholes worked satisfactorily with steady state flow and 2 m intervals for hydraulic conductivities as low as 10^{-11} m/s. Increasing the test interval to 20 m would have enabled an evaluation of conductivities as low as 10^{-12} m/s, but the longer interval would have been many times greater than the average fracture spacing. This was not desirable because it would have decreased the statistical information on aperture distributions. Pressure pulse techniques applied to 2 m intervals could decrease this limit down to 10^{-12} m/s and lower, but the application of this method to fractured rocks needs extensive evaluation [15].

The new technique of measuring the average rock mass permeability that was obtained from the macropermeability experiment can also be improved. The final result of about 10^{-10} m/s could be decreased by one order simply by making the drift ten times longer. By improving the method of detecting low flow rates, it should be possible to detect seepage rates as low as 5 ml/min and as the depth of the test chamber increases, the effective hydraulic gradients could be 5-10 times greater than those measured at Stripa. Thus, hydraulic conductivities of the order of 10^{-12} m/s should be measurable within reasonable time periods.

A comparison of the borehole results with those of the macropermeability experiment provides a unique opportunity to evaluate two different methods of measuring fractured rock permeabilities. The average permeability of 10^{-13} cm^2 from the macropermeability experiment compares quite favorably with the average of 8.9×10^{-13} cm^2 determined from borehole testing in the rock walls of the same room. The higher average value determined with the latter method probably reflects some differences in test mode. For example, testing in boreholes required fluid injection, whereas fluid was being withdrawn from the system in the large scale experiment. Also, the lengths of individual boreholes that radiated out in all directions increased the possibility that more high permeability fractures would be intersected than by the horizontal drift. The macropermeabilty method, however, appears to be the one technique that could be used during development of actual repository sites to determine the acceptability of individual storage drifts.

Although the procedures that will be used to characterize a repository site have yet to be decided, it seems likely that they will include some form of probability analysis. This will require that the pertinent properties of the rock mass be described in terms of appropriate statistical distributions. The borehole testing program has demonstrated how data on the variation in fracture permeabilities can be obtained. Ultimately, one needs to be able to determine rate of movement of radionuclides through the rock mass, and this will require data on fracture apertures. The testing program at Stripa has demonstrated how different interpretational models can be used to compute statistical distributions for fracture apertures. As discussed above, we are now investigating a new method of computing fracture porosity, which of course is necessary in determining true velocities. In addition, when fracture apertures are combined with fracture orientations and spacings, it will be possible to determine directional permeabilities. The macropermeability method, which yields an average rock mass permeability for a very large volume of rock, will provide a necessary check on the results calculated from borehole data.

The approach taken to develop the fracture hydrology data base at Stripa will also enable one to evaluate the degree of hydraulic anisotropy within the rock mass. This can then be used to determine the distribution of velocities within the fracture system and the most probable fracture pathways for radionuclide migration. This level of understanding for the fracture hydrology should also be a necessary prerequisite to the planning and execution of any tracer studies. Completion of the originally planned tracer experiments would permit a comparison of predicted versus measured transit times and thus serve as a check on the validity of our interpretations of the fracture hydrology at Stripa.

A more complete understanding of the effective transit time within the fractured rock mass at Stirpa is necessary if we are to properly interpret the apparent groundwater ages determined by analysis of the isotopes and the groundwater residence times. This will certainly require a thorough analysis of fracture porosity and permeability. It is obvious, given the complexity of the highly disturbed flow system at Stripa, that a necessary objective of the fracture hydrology program should be an attempt to reconcile the measured porosities, permeabilities and flow system boundary conditions with the distribution of isotopes and groundwater chemistry. Such an attempt, even if only partially successful, will provide considerable insight into the nature of flow systems in fractured rocks and should enhance the overall problem of selecting sites for nuclear waste disposal.

Finally, the hydrogeological investigations at Stripa have demonstrated the value of coordinated surface and underground testing in site characterization. It is recognized that the logical sequence proceeds from surface work to the subsurface with appropriate analysis of results at key points in the total process. The hydrological as well as the thermomechanical behavior of the rock mass is controlled by fractures and other discontinuities, and predictions made using standard continuum assumptions are often not valid. Proper evaluation of rock mass properties will require detailed studies of the influence of these discontinuities, and will involve adequate sampling of their geometric and hydraulic properties from both surface-based and at-depth tests. The work at Stripa has shown that surface and underground testing provide complementary information, and that the complete suite of needed data cannot be obtained from one approach alone.

ACKNOWLEDGMENTS

The results that are summarized here represent a cumulative effort that had its beginnings during a workshop on "Movement of Fluids in Largely Impermeable Rock" that was held January 27-29, 1977, in Austin Texas under the sponsorship of the Office of Waste Isolation (OWI), U.S. Energy and Development Administration (now part of the U.S. Department of Energy). The workshop was organized at the request of Dr. T.F. Lomenick, who was then with OWI. Subsequent to that meeting he played a key role in setting up the initial organization of the Stripa project. We acknowledge with thanks the valuable assistance rendered to this project by Dr. Lomenick. The present work was supported by the Assistant Secretary for Nuclear Energy, Office of Waste Isolation of the U.S. Department of Energy under contract DE-AC03-76SF00098. Funding for this project is administered by the Office of Nuclear Waste Isolation at Battelle Memorial Institute.

REFERENCES

1. Witherspoon, P.A. and Degerman, O.: "Swedish-American Cooperative Project on Radioactive Waste Storage in Mined Caverns," Lawrence Berkeley Lab. Rep. LBL-7049, SAC-01, 1978.

2. Gale, J.E. and Witherspoon, P.A.: "An Approach to the Fracture Hydrology at Stripa, Preliminary Results," Lawrence Berkeley Lab. Rep. LBL-7079, SAC-15, 1979.

3. Olkiewicz, A., Gale, J.E., Thorpe, R. and Paulsson, B.: "Geology and Fracture System at Stripa," Lawrence Berkeley Lab. Rep. LBL-8907, SAC-21, 1979.

4. Wollenberg, H., Andersson, L. and Flexser, S.: "Petrology and Radiogeology of the Stripa Pluton," Lawrence Berkeley Lab. Rep. LBL-11654, SAC-36 (in preparation), 1982.

5. Thorpe, R.: "Characterization of Discontinuities in the Stripa Granite Time-Scale Heater Experiment," Lawrence Berkeley Lab. Rep. LBL-7083, SAC-20, 1979.

6. Paulsson, B.N.P., Nelson, P.H. and Kurfurst, P.J.: "Characterization of Discontinuities in the Stripa Granite - Full Scale Heater Experiments," Lawrence Berkeley Lab. Rep. LBL-9063, (in preparation), 1982.

7. Rouleau, A., Gale, J.E. and Baleshta, J.: "Results of Fracture Mapping in the Ventilation Drift - Stripa," Lawrence Berkeley Lab. Rep. LBL-13071 (in preparation), 1982.

8. Wilson, C.R., Long, J.C.S., Galbraith, R.M., Karasaki, K., Endo, H.K., DuBois, A.O., McPherson, M.J. and Ramquist, G.: "Geohydrological Data from the Macropermeability Experiment at Stripa," Lawrence Berkeley Lab. Rep. LBL-12520 (in preparation), 1982.

9. Wilson, C.R., Witherspoon, P.A., Long, J.C.S., Galbraith, R.M., DuBois, A.O. and McPherson, M.J.: "Large Scale Conductivity Measurements from Radial Flow into an Underground Opening in Fractured Granite," Lawrence Berkeley Lab. Rept. LBL-14876 (in preparation), 1982.

10. Gale, J.E.: "Fracture and Hydrology Data from Field Studies at Stripa, Sweden," Lawrence Berkeley Lab. Rep. LBL-13101 (in preparation), 1982.

11. Rouleau, A. and Gale, J.E.: "Characterization of the Fracture System at Stripa," Lawrence Berkeley Lab. Rep. LBL-14875 (in preparation), 1982.

12. Baecher, G.B., Lanney, N.A. and Nicholas, A.: "Trace Length Biases in Joint Surveys," Proc. 19th U.S. Rock Mechanics Symposium, pp. 56-65, 1978.

13. Rouleau, A. and Gale, J.E.: "Characterizing Fracture Systems for Hydrogeological Purposes - Application to the Gneissic Bedrock at Chalk River," Atomic Energy of Canada, Tech. Rep. (in press). 1982.

14. Baecher, G.B.: "Progressively Censored Sampling of Rock Joint Traces," Mathematical Geology, 12 (1), pp. 133-137, 1980.

15. Gale, J.E.: "Hydrogeologic Characteristics of the Stripa Site," Lawrence Berkeley Lab. Rep. LBL-14878 (in preparation), 1982

16. Knapp, R.B.: "An Analysis of the Porosities of Fractured Crystalline Rocks," Univ. Arizona, M.S. Thesis, 90 p., 1975.

17. Withespoon, P.A., Nelson, P., Doe, T., Thorpe, R., Paulsson, B., Gale, J., and Forster, C.: "Rock Mass Characterization for Storage of Nuclear Waste in Granite," Lawrence Berkeley Lab. Rep. LBL-8570, SAC-18, 1979.

18. Fritz, P., Barker, J.F. and Gale, J.E.: "Geochemistry and Isotope Hydrology of Groundwaters in the Stripa Granite," Lawrence Berkeley Lab. Rep. LBL-8285, SAC-12, 1979.

19. Fritz, P., Barker, J.F. and Gale, J.E.: "Summary of Geochemical Activities at the Stripa Site During FY79/80," Lawrence Berkeley Lab. Rep. LBL-11864 (in preparation), 1982.

20. Marine, I.W.: "Geochemistry of Groundwater at the Savannah River Plant," Dupont de Nemours and Co. Rep. No. DP 1356, 1976.

21. Witherspoon, P.A., Wilson, C.R., Long, J.C.S., Galbraith, R.M., DuBois, A.O. Gale, J.E. and McPherson, M.J.: "Mesures de perméabilite en grand dans les roches cristallines fracturées," Bull. B.R.G.M., III (1), pp. 53-61, 1980-81.

22. Long, J.C.S., Witherspoon, P.A., Wilson, C.R. and DuBois, A.O.: "Large-Scale Permeability Testing at Stripa," Proc. Third Invitational Well-Testing in Low Permeability Environments, Lawrence Berkeley Lab. Rep. LBL-12076, 1980.

23. Long, J.C.S., Remer, J.S., Wilson, C.R. and Witherspoon, P.A.: "Porous Media Equivalents for Networks of Discontinuous Fractures," Water Resources Res. 18 (3), pp. 645-658, 1982.

INTERPRETATION OF FIELD EXPERIMENTS ON THE FLOW OF WATER AND TRACERS THROUGH CRYSTALLINE ROCK

D.P. Hodgkinson, D.A. Lever, P.C. Robinson and P.J. Bourke
Theoretical Physics and Chemical Technology Divisions
AERE Harwell, Oxfordshire

ABSTRACT

This paper reviews recent work at Harwell on the interpretation of field experiments on the flow of water and tracers through crystalline rock. First a model for the radial transport of tracers through an isolated fracture is outlined and used to analyse a recent Swedish experiment at Finnsjön. Secondly, the theoretical and experimental approach that is being used to quantify flow and dispersion through networks of fractures is described.

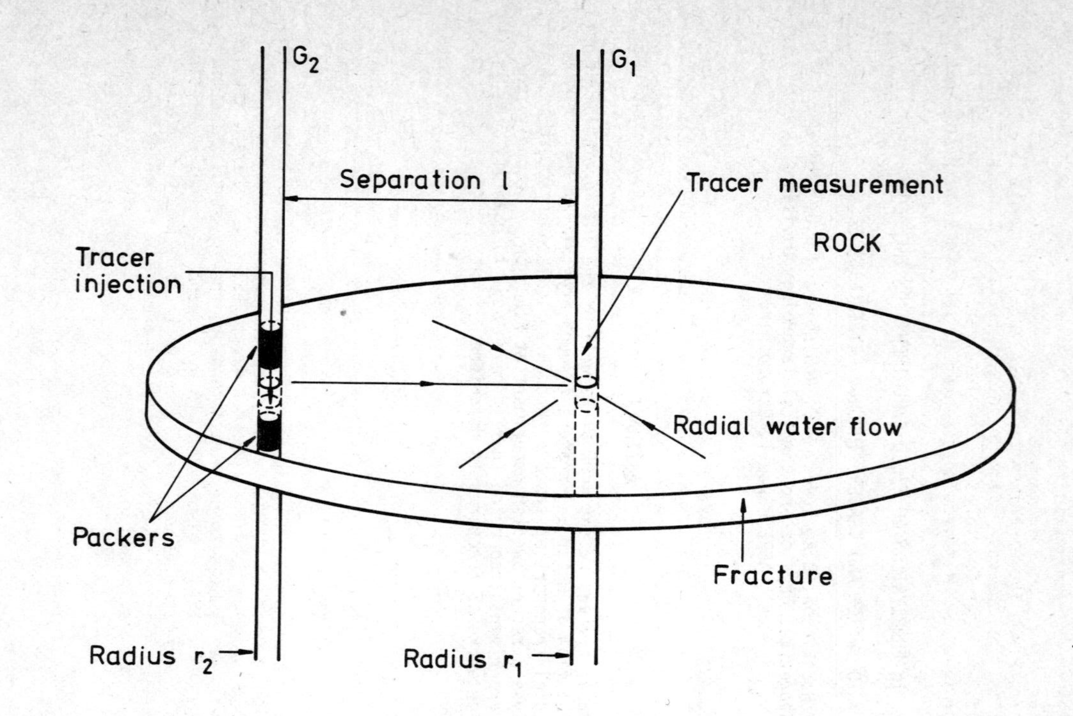

FIG.1.CONCEPTUAL MODEL OF THE EXPERIMENT

1. Introduction

Safety assessments of radioactive waste burial require extensive use of mathematical models for the transport of radionuclides in groundwater. If reliable predictions are to be made then these models must be capable of interpreting field experiments on water flow and tracer movement through geological formations. This paper reviews recent work, carried out by AERE Harwell, on the interpretation of such experiments in crystalline rock.

It is clear from the experiments carried out in Cornwall [1, 2] and elsewhere [3-7] that most of the water flow in such rocks takes place through a network of approximately planar fractures. The intervening rock is much less permeable to water flow but dissolved species can diffuse into its micropores under the action of a concentration gradient [8-13]. In addition to advection and rock matrix diffusion, the migration of tracers can be affected by chemical sorption on to the rock and hydrodynamic dispersion both within individual fractures and as a result of mixing at fracture intersections.

The relative importance of these various mechanisms is examined for isolated fractures in section 2 and for fracture networks in section 3.

2. Isolated fractures

Many field experiments have been performed on the flow of water and tracers through fracture zones in crystalline rock. In general, both inter-hole pressure drop measurements and tracer breakthrough curves are difficult to analyse. This is partly due to the fact that fractures are rarely sufficiently isolated from one another; if they were truly isolated then there would be no interconnected permeability. Thus in many experiments there are indeterminable perturbations to the pressure field, extra unidentified migration paths and poorly characterised boundary conditions.

However, one experiment on an isolated fracture [14, 15] was considered by the International Nuclide Transport Code Intercomparison Study (INTRACOIN) [16] to be suitable for detailed analysis by the participating groups. A schematic view of the experiment, which was performed by Gustafsson and Klockars [14] at Finnsjön in Sweden, is shown in Figure 1. Sorbed and non-sorbed tracers (Sr^{2+} and I^-) were injected for a period of 350 hours into borehole G_2 between two packers straddling a fracture zone at a depth of about 100m. Water was continuously pumped out of the measurement borehole (G_1) with the aim of creating a drawdown zone into which all the tracers would eventually flow.

This experiment has been analysed with a model which includes the effects of radial advection, hydrodynamic dispersion, kinetic sorption and diffusion into the rock matrix [17]. The model has several novel features. First, the longitudinal dispersivity is assumed to be proportional to the square of the mean radial velocity by analogy with Taylor dispersion [18]. Secondly, kinetic sorption on to the fracture filling material and the altered rock close to the fracture surface is included. This is modelled by a first-order kinetic reaction on to the surface of a plane-parallel fracture. However, it is interpreted as a phenomenological model for sorption in the rather ill-defined region in the vicinity of the fracture. Thirdly, a special outlet boundary condition is used in order to model the conditions of the present experiment as closely as possible. Finally, the model is evaluated by numerical inversion of the analytical solution to the Laplace transformed equations [19, 20]. This has proved to be a very versatile, efficient and accurate technique for the solution of this type of problem.

The best fit to the I^- data is shown as a continuous line in Figure 2. It is seen that the data is very well fitted by the model. In particular, the tail at times greater than about 375 hours is reproduced well. This is due to ions diffusing out of the rock giving a concentration which falls off as $t^{-3/2}$ [21]. For the present parameters hydrodynamic dispersion is unimportant in comparison with rock matrix diffusion. This is illustrated by the dashed line in Figure 2 which is calculated for the same parameters as before except that the diffusion term is omitted. The experimental recovery of I^- was 98% after 600 hours compared to the fitted value of 95%.

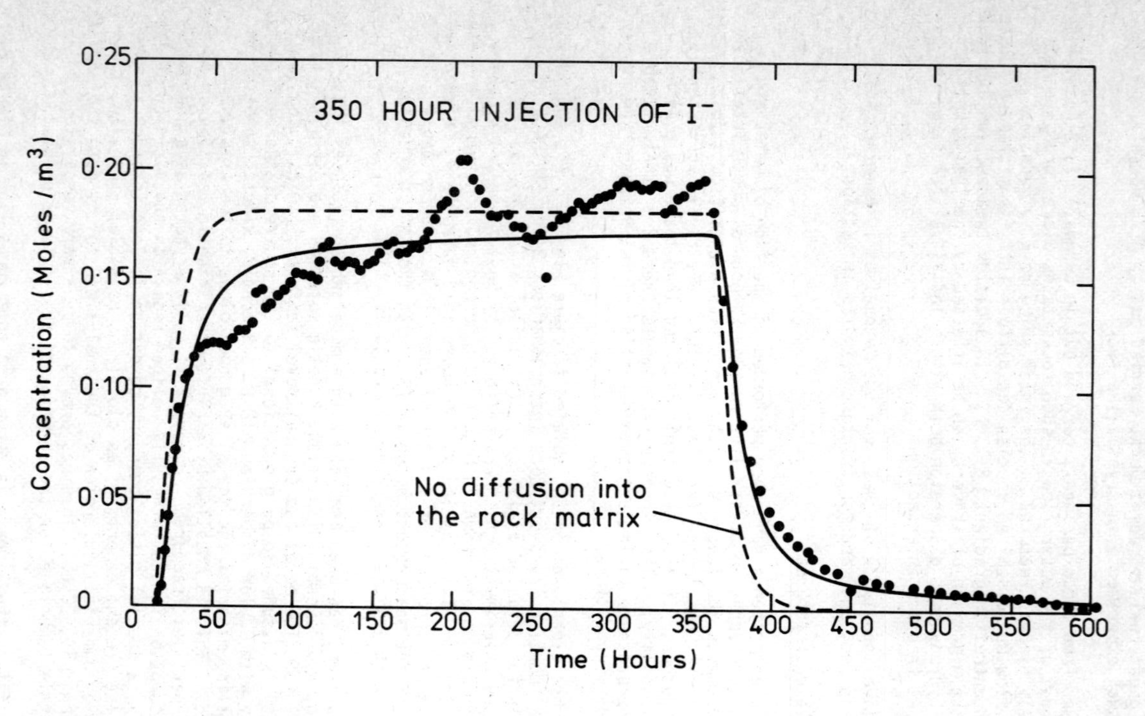

FIG.2. CONCENTRATION IN THE MEASUREMENT BOREHOLE AS A FUNCTION OF TIME FOR A 350 HOUR INJECTION OF I^- (RUN 2). THE CONTINUOUS LINE IS THE BEST FIT TO THE DATA WHILE THE DASHED LINE SHOWS THE EFFECT OF OMITTING DIFFUSION INTO THE ROCK MATRIX.

The recovery of Sr^{2+}, however, was only 60%. Thus the crucial question here is to understand and quantify the loss mechanism. It was not possible to fit the data with an equilibrium sorption model. Instead, it is assumed that the loss is due to the kinetics of sorption on to the fracture filling material and the degraded rock near the fracture surface. Moreover, as 40% of the tracer is lost, this kinetic sorption must be effectively irreversible during the period of the experiment. With this assumption, an excellent fit to the data was obtained with one diffusion parameter and one kinetic sorption rate in addition to the two hydrogeological parameters determined from the non-sorbed I^- data. The best fit is shown as a continuous line in Figure 3. Once again the importance of out diffusion in fitting the tail is seen by comparing with the dashed line in Figure 3 for which there is no diffusion into the rock matrix. The best fit gave a 61% recovery of Sr^{2+} after 600 hours compared to the experimentally measured 60%.

During the course of the experiment the tracers diffuse at most a few centimetres into the rock. This rock will therefore mainly consist of fracture filling material and degraded rock near the fracture surface. Thus the in-situ diffusivities found from the above analysis are not expected to be the same as those measured for intact samples of rock in laboratory experiments. In fact the in-situ intrinsic (or effective) diffusivity determined here is found to be between one and four orders of magnitude larger than the values found in such laboratory experiments [8, 11]. It would be interesting to try and measure the dispersion and sorption parameters of rock in and around a water bearing fracture for comparison with the in-situ values derived from the present analysis.

In future experiments of this type it would be interesting to considerably increase the period of injection (e.g. to a year) in order that tracer has time to diffuse into the unaltered rock. The extent of this diffusion could then be checked by coring perpendicular to the fracture.

3. Fracture networks

In modelling radionuclide migration through crystalline rock masses it is necessary to take into account the properties of the fracture network to some order of approximation. The first-order approximation is to use a continuum description of the rock mass which essentially averages over a region containing a large number of intersecting fractures. For problems which only involve the flow of water this approach is simply the familiar permeable medium model which is characterised by a permeability and a porosity. Continuum models can also be formulated which include the effects of sorption, rock matrix diffusion [21] and macroscopic hydrodynamic dispersion [22]. However, the conditions under which these continuum migration models adequately approximate a fracture network are not well established. Clearly they are only valid on a length scale which is large compared to the fracture dimensions. However, there may also be other limitations on their validity. For instance, the Fickian approximation conventionally used for modelling hydrodynamic dispersion due to mixing at fracture intersections may only be valid at very large times [23-25].

Given this uncertainty in the applicability of continuum migration models, it is worthwhile trying to model the flow through fracture networks more explicitly. A completely deterministic fracture flow model would require a detailed knowledge of the positions, orientations, lengths and widths of all fractures in the region of interest. Detailed information on these quantities is unlikely to become available. Its collection would probably be prohibitively expensive and destroy the site being investigated. It is also unnecessarily detailed given the large number of uncertainties in modelling radionuclide migration.

The alternative approach being followed at Harwell [26, 27] and elsewhere [23-25, 28-30] is to incorporate the essential features of the fracture system into a model in a probabilistic rather than a deterministic way. The data requirements are thereby reduced to manageable proportions.

The basic assumptions of the model are that fractures are planar, of finite extent, occur randomly in the rock mass and have the following characteristics which are described by uncorrelated probability distributions:

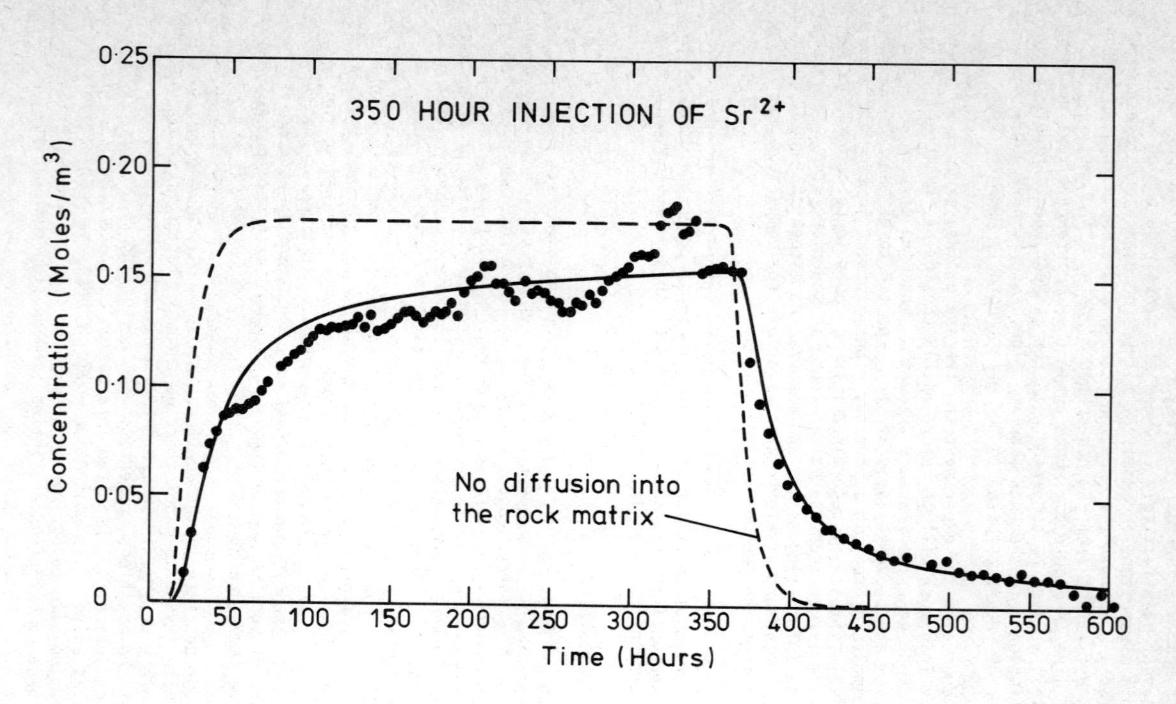

FIG.3. CONCENTRATION IN THE MEASUREMENT BOREHOLE AS A FUNCTION OF TIME FOR A 350 HOUR INJECTION OF Sr^{2+} (RUN 2). THE CONTINUOUS LINE IS THE BEST FIT TO THE DATA WHILE THE DASHED LINE SHOWS THE EFFECT OF OMITTING DIFFUSION INTO THE ROCK MATRIX.

(i) Number of fractures per unit volume,

(ii) Orientation,

(iii) Length,

(iv) Width.

Given this limited statistical information it is possible to construct a computer model of the fracture system and the flow through it. Fractures with appropriate orientations, lengths and widths are generated by random sampling of the probability distributions until the required fracture density is reached. In this way it is possible to generate a large number of statistically equivalent fracture networks through which the flow of water and tracers can be investigated.

One application of this approach is to study the validity of continuum models. Also it can be used to aid the interpretation of experimental data, to mimic field experiments, and to provide permeability values for input to a regional flow model. It is currently being used to examine the connectivity, water flow and hydrodynamic dispersion for some idealised probability distributions in two dimensions, although some results on connectivity in three dimensions have been obtained [26, 27]. Recent progress on these topics is discussed briefly below.

3.1 Connectivity

It is often taken for granted that fractured rock will be open to bulk flow. However, the existence of flow paths depends on the fractures being sufficiently interconnected. For a given set of fracture statistics it is found that as the fracture frequency is increased the probability that a connected flow path exists right across the network jumps quite suddenly from zero to one. In fact in an infinite system this jump occurs at a single critical value of the fracture frequency. This critical frequency obviously depends on the fracture statistics but another measure of the interconnectedness is found to be largely independent of these statistics. This is the average number of intersections per fracture. In two dimensions this number is always found to lie between 3 and 4 at the point when connected flow paths appear [26]. In three dimensions it appears that the number is lower being around 2.2 intersections per fracture.

3.2 Flow

The flow through model fracture networks arising from applied pressure gradients has been calculated and used to deduce the permeability tensor [27]. This has been repeated for different pieces of rock with the same statistics in order to find the variability and size-dependence of the permeability. For example, in one particular case the variability in permeability for different pieces of rock became quite small for pieces which were about ten times the fracture length. Also, the size-dependence of the permeability was no longer significant at this length scale, which therefore represents the lower limit of applicability for a continuum approximation for this example. However, the statistics for this example had very little variability. It is expected that the size required to obtain a constant permeability will be somewhat larger for more realistic cases with larger statistical variations.

3.3 Hydrodynamic dispersion

Once the flow has been calculated, the movement of tracers along the fractures can be followed. At fracture intersections it is assumed that complete mixing occurs so that a pulse of tracer spreads out as it flows through the system. At present other effects, such as sorption and diffusion into the rock matrix, are not included although their inclusion is not an insurmountable problem. The model is being used to investigate the validity of the Fickian diffusion-like approximation to hydrodynamic dispersion used in conventional continuum models.

3.4 Field experiments

The above network modelling requires experimental information on the distribution of orientations, lengths and widths of water bearing fractures and their frequency of occurrence. Experimental techniques have been developed at a granitic field site in Cornwall in order to try and obtain this information [2].

Conventional packer measurements with short (about 1 m) pressurised zones, core-logging and geophysical tests have been used to determine the frequency of water bearing fractures. Such measurements have been made in several holes down to 200m and are continuing down to 800m in one hole. The results obtained so far show that more than 90% of the water injected enters 30 discrete fractures which occur randomly and have an average 10m separation along the holes.

Radioactive tracers are then pumped into single fractures and detectors in adjacent holes determine the positions of single or multiple appearances of tracers. These results provide orientation and topological information about the fracture intersections. To date, tracers have been injected into about 12 fractures in three holes and tracer arrival has been observed in adjacent holes with radial separations of between 5m and 30m from the injection holes. About 40 arrival positions have been found, again with an average separation of about 10m.

Finally, pairs of multi-packer assemblies are set at positions shown by tracers to be on the same fracture and interhole flow and pressure drop tests are made to obtain estimates of the effective hydraulic and geometric apertures.

The present experimental programme will not give any direct measure of the length of individual fractures. However, the tracer tests give some information about the occurrence of intersections between fractures which, together with a knowledge of the average distance between fractures, can be used to give an indirect estimate of their length.

The experimental approach outlined above is being pursued rather than the more conventional approach of deducing permeabilities from pumping tests in single boreholes, for the following reasons. In single hole injection tests with the hole diameter inevitably much smaller than the 10m separation between fracture intersections, the flows measured are determined mainly by the fracture aperture locally round the hole and independent of the interconnections of the pattern. Inter-hole pressure response and tracer measurements which would have to be over many tens of metres for direct measurements of permeability do not seem practical below the highly fractured top 50m of the Cornish site.

A further point arises from the importance of the retardation and dispersion of radionuclides. These processes depend on the thickness of rock between flow paths and the distances between intersections of paths, which are measured quantities in the present experimental approach.

Acknowledgement

This work has been commissioned by the Department of the Environment, as part of its radioactive waste management research programme. The results will be used in the formulation of Government policy, but at this stage they do not necessarily represent Government policy.

The work was performed under contract with the European Atomic Energy Community in the framework of its R & D programme on Management and Storage of Radioactive Waste.

References

1. Bourke, P.J., Bromley, A., Rae, J., and Sincock, K.: "A Multi-Packer Technique for Investigating Resistance to Flow through Fractured Rock and Illustrative Results", Proc. Workshop on Siting of Radioactive Waste Repositories, 173-190, OECD/NEA, Paris, May 1981.

2. Bourke, P.J., Evans, G.V., Hodgkinson, D.P. and Ivanovich, M.: "An Approach to Prediction of Water Flow and Radionuclide Transport through Fractured Rock", Proc. Workshop on the Investigation of Rock for Burial of Radioactive Waste, OECD/NEA, Ottawa, September 1982.

3. Witherspoon, P.A., Cook, N.G.W., and Gale, J.E.: "Geologic Storage of Radioactive Waste : Field Studies in Sweden", Science, Vol. 211, 894-900, 1981.

4. Bertrand, L., Breton, J-P., Genetier, B., and Vaubourg, P., "Reconnaissance d'un Massif Rocheux à Grande Profondeur Mesure de L'Orientation des Carottes de Forage Mesure de Faibles Valeurs de Perméabilité", Proc. Workshop on Siting of Radioactive Waste Repositories, 235-247, OECD/NEA, Paris, May 1981.

5. Davison, C.C., "Physical Hydrogeology Measurements Conducted in Boreholes WN-1, WN-2 and WN-4 to Assess the Local Hydraulic Conductivity and Hydraulic Potential of a Granitic Rock Mass", AECL-TR-26, 1980.

6. Davison, C.C., "Physical Hydrogeologic Measurements in Fractured Crystalline Rock - Summary of 1979 Research Programs at WNRE and CNRL", AECL-TR-161, 1981.

7. Raven, K.G., and Smedley, J.A., "CNRL Groundwater Flow Study - Summary of FY 1981 Research Activities", Environment Canada, NHRI-INRH, 1982.

8. Bradbury, M.H., Lever, D.A., and Kinsey, D.V., "Aqueous Phase Diffusion in Crystalline Rock", AERE-R.10525, Proc. Fifth International Symposium on the Scientific Basis for Radioactive Waste Management, Berlin, June, 1982.

9. Hemingway, S.J., Bradbury, M.H., and Lever, D.A., "The Effect of Dead-End Porosity on Rock Matrix Diffusion", AERE-R.10691, 1982.

10. Bradbury, M.H., and Lever, D.A., "A proposal for a Laboratory-Scale Migration Experiment", AERE-R.10667, 1982.

11. Skagius, K., and Neretnieks, I., "Diffusion in Crystalline Rocks", Proc. Fifth International Symposium on the Scientific Basis for Radioactive Waste Management, Berlin, June, 1982.

12. Wadden, M.M., and Katsube, T.J., "Radionuclide Diffusion Rates in Igneous Crystalline Rock", Proc. Joint Annual Meeting of GAC-CGU, May, 1981.

13. Neretnieks, I., "Diffusion in the Rock Matrix : An Important Factor in Radionuclide Retardation?", J. Geophys. Res., 85, 4379, 1980.

14. Gustafsson, E., and Klockars, C-E., "Studies of Groundwater Transport in Fractured Crystalline Rock under Controlled Conditions using Nonradioactive Tracers", KBS 81-07, 1981.

15. Davison, C., Goblet, P., and Neretnieks, I., "Tracer Tests in Fissured Rock for Model Testing in INTRACOIN; Appendix 1, Introductory Description of tracer tests at Finnsjön and Summary of Important Data", 1982.

16. INTRACOIN progress reports, No. 1 (1981), No. 2 and Nos. 3 (1982), Swedish Nuclear Power Inspectorate, Box 27106, 102 52 Stockholm, Sweden.

17. Hodgkinson, D.P., and Lever, D.A., "Interpretation of a Field Experiment on the Transport of Sorbed and Non-sorbed Tracers Through a Fracture in Crystalline Rock", AERE-R.10702, 1982.

18. Taylor, G.I., "Dispersion of Soluble Matter in Solvent Flowing Slowly through a Tube", Proc. Roy. Soc. A219, 186, 1953.

19. Barker, J.A., "Laplace Transform Solutions for Solute Transport in Fissured Aquifers", Adv. Water Resources, 5, 98, 1982.

20. Talbot, A., "The Accurate Numerical Inversion of Laplace Transforms", J. Inst. Math. Appl., 23, 97, 1979.

21. Lever, D.A., Bradbury, M.H., and Hemingway, S.J., "Modelling the Effect of Diffusion into the Rock Matrix on Radionuclide Migration", AERE-R.10614, 1982.

22. Scheidegger, A.E., The Physics of Flow Through Porous Media, 3rd edn., University of Toronto Press, 1974.

23. Schwartz, F.W., "Macroscopic Dispersion in Porous Media : The Controlling Factors", Water Resources Research, Vol. 13, 743-52, 1977.

24. Smith, L., and Schwartz, F.W., "Mass Transport 1. A Stochastic Analysis of Macroscopic Dispersion", Water Resources Research, Vol. 16, 303-13, 1980.

25. Schwartz, F.W., Smith, L., and Crowe, A.S., "Stochastic Analysis of Groundwater Flow and Contaminant Transport in a Fractured Rock System", preprint, 1982.

26. Robinson, P.C., "Connectivity of Fracture Systems - A Percolation Theory Approach", UKAEA Report TP.918, 1982, J. Phys. A : Math. Gen. (to be published), 1982.

27. Robinson, P.C., "NAMNET - Network Flow Program", AERE-R.10510, 1982.

28. Smith, L., and Schwartz, F.W., "Mass Transport 2. Analysis of Uncertainty in Prediction", Water Resources Research, Vol. 17, 351-69, 1981.

29. Smith, L., and Schwartz, F.W., "Mass Transport 3. Role of Hydraulic Conductivity Data in Prediction", Water Resources Reseach, Vol. 17, 1463-79, 1981.

30. Long, J.C.S., Remer, J.S., Wilson, C.R., and Witherspoon, P.A., "Porous Media Equivalents for Networks of Discontinuous Factures", Water Resources Research, Vol. 18, 645-58, 1982.

HYDROGEOLOGICAL INVESTIGATIONS IN BOREHOLES

L. Carlsson*, H. Norlander** and T. Olsson***
* SGAB, Göteborg, Sweden
** SMS, Stripa, Sweden
***K-Konsult, Stockholm, Sweden

ABSTRACT

Hydrogeological investigations in boreholes is one of the programs included in the current Stripa Project. The aim with the program is to develop methods and instruments for hydrogeological and geochemical investigations in subsurface boreholes. The aim is also to obtain further data and knowledge on deep-lying bedrock and groundwater. The investigation program includes geophysical logging, hydraulic tests, rock stress measurements, hydrogeochemical sampling and analyses. The investigations are carried out in four boreholes, two vertical (505 and 822 m in length) and two subhorizontal holes (each 300 m in length) drilled at two test sites down in the Stripa mine. The boreholes penetrate the Stripa granite to their full lengths. Generally, the rock mass is moderately fractured, but a highly fractured zone has been found in one of the vertical holes. The configuration of this zone was investigated in part by means of cross-hole techniques. Water samples were taken for detailed chemical and isotopic analyses.

Underground testing methods enable investigations to be carried out of structures and elements of interest and importance on the actual level for repository construction. The investigations performed support this claim and also show that accurate data and information can be obtained from horizontal boreholes under existing groundwater conditions.

1 BACKGROUND

1.1 General

The final planning and layout of a deep underground repository for nuclear waste or spent fuel requires detailed information on geological and hydrogeological conditions at repository depth. For many reasons, it is impractical to acquire this information by investigations from the surface alone. Underground investigation methods for this information must therefore be developed and tested,and the collecting validity of their results must be demonstrated.

Underground drilling of horizontal or subhorizontal boreholes is a key element in such investigations prior to actual tunnelling. Such boreholes permit accurate mapping of rocks, structures and groundwater conditions in the penetrated sections, as well as sampling and testing of the rocks and the local fluids and gases. In addition, the rock volumes surrounding the boreholes can be tested by a variety of single- and cross-hole techniques.

Hydrogeological investigations in boreholes is one of the main programs included in the current phase of the Stripa Project, where various techniques and tools are to be developed and tested at depth. The purposes of the program are as follows.

1. Development of methodology for hydrogeological and hydrogeochemical investigations in subsurface horizontal and vertical boreholes.

2. Development of instrumentation and equipment for use in subsurface horizontal and vertical boreholes.

3. Hydraulic, chemical and isotopic characterization of the Stripa granite and groundwaters.

The work is being carried out in accordance with the defined program /1/, which was slightly revised during its performance. At present, the program includes drilling, core-logging, geophysical logging, hydraulic investigations, rock stress measurements, water sampling and water analyses.

1.2 Test sites

The program is being carried out in boreholes at two specific sites down in the Stripa mine. At the SGU site, which is the main site located at the 360 m mine level, three boreholes were drilled, one vertical (V1) and two subhorizontal boreholes (N1 and E1). A fourth borehole was drilled from a second site at the 410 m level. This borehole is the old Dbh V1 borehole drilled during the Swedish-American Cooperation (SAC) program and now deepened to a total depth of 822 m (1230 m below the ground surface). The locations of the sites and boreholes are shown in Figure 1.

1.3 Geology

The target rock, the Stripa granite, is a grey to reddish, medium-grained granitic rock type of Precambrian age (1691 My). On the surface, it is located north of the mine and is limited in extent towards the south by the orebearing series of metamorphic rock. The contact strikes approximately E-W and is almost vertical near the surface. At greater depth, it swings towards the south, and at a depth of about 400 m, as exposed in the mine, it is nearly horizontal. The general strike of the host metamorphic rocks (leptites) suggests that the granite may be elongated in an E-W direction. It may possibly represent the apex of a larger intrusion, which widens at depth, as indicated by the southern contact.

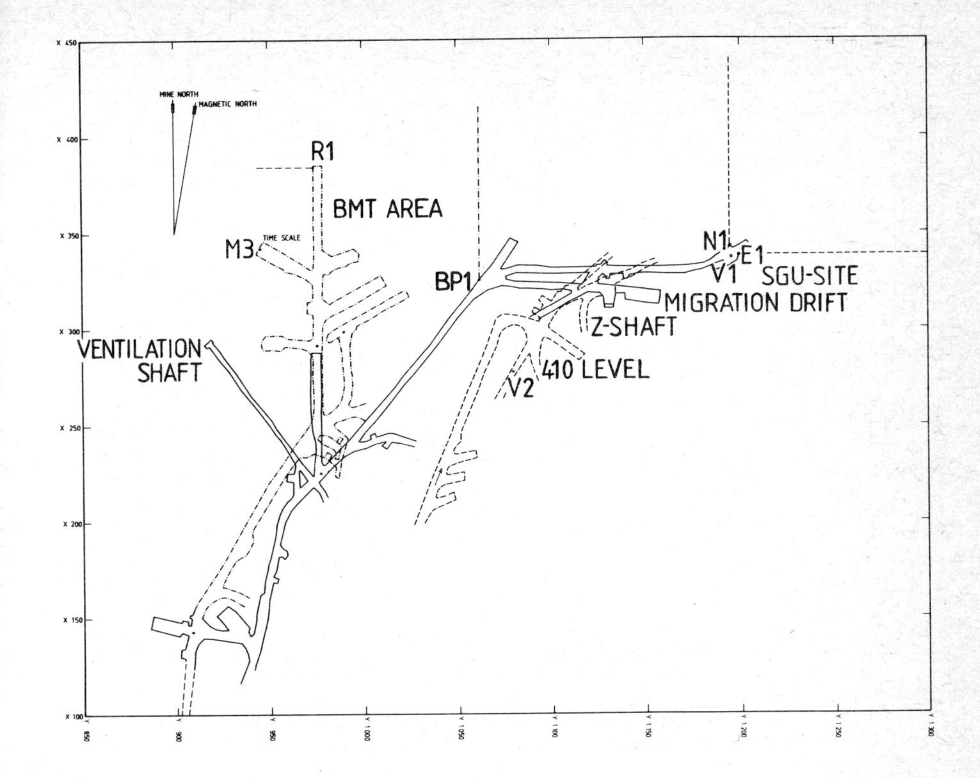

Figure 1. Sites and boreholes in the Stripa Mine.

As regards the lineaments in the granite, their directions are generally parallel to the syncline axis of the supracrustal formation. There are two dominant directions for the lineaments, ENE-WSW and NW-SE.

A number of zones of fractured or crushed rock exist in the granite. Normally these zones are thin, not exceeding 1 m in the cores, but a few zones are several meters in thickness. A more extensive fracture zone was found in the lowermost part of borehole V1. In the upper 466 m of the borehole, the granite is slightly tectonized, and contains widely spaced fracture and crush zones usually less than 1 m in width. Fracturing tends to be more intense towards the bottom of this section, with a marked increase in the number of subvertical fractures. Below 466 m and at the bottom of the borehole (505 m) as much as 7.7 m of the core is disintegrated (cf. Table 1).

The fracture zone in V1 has a high rate of water inflow. Hydraulic conductivity is high in comparison to the rock mass and the zone is of crucial importance for the presence and flow of groundwater in the granite. However, different explanations may be given for the extent and orientation of the zone.

In principle, any of the following interpretations, singly or in combination, can explain the configuration of the fracture zone found in V1:

* A zone striking N70E and dipping 60SE, indicated by existing zones found in other parts of the mine.

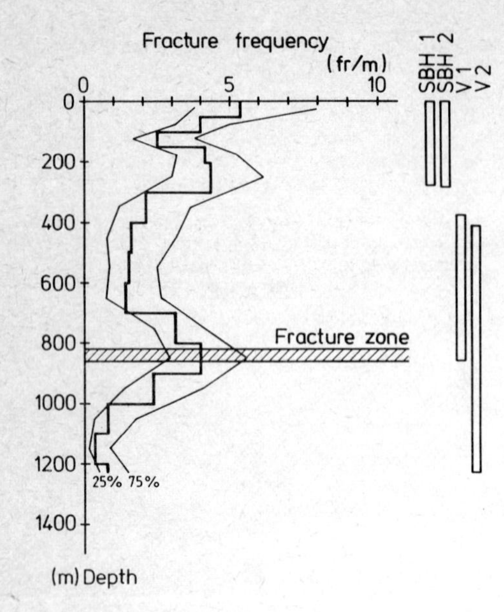

Figure 2. Fracture frequency versus depth in Stripa calculated from the SBH-1, SBH-2, V1 and V2 boreholes. Frequency is assumed to have a log-normal distribution and the 25 %, median and 75 % values are given.

* A zone striking NW-SE steeply dipping NE, indicated by existing zones found in other parts of the mine.

* The zone is divided into several minor zones further away from the borehole, as indicated by geophysical measurements.

The degree of fracturing of the Stripa granite varies considerably. Based on data obtained from both surface and subsurface boreholes, the variation of fracture frequency versus depth is shown in Figure 2. Table 1 presents the mean fracture frequencies obtained in the boreholes included in the hydrogeological program.

Table 1. Fracture frequency in V1, V2, N1 and E1.

Borehole	Fracture frequency (fr/m)
V1 (above the crush zone)	1.5
V1 (crush zone)	12.9
V2	2.1
N1	1.6
E1	4.7

Steep fractures dominate the fracture system in the Stripa granite, and make up as much as 49 % of all fractures in the rock mass adjacent to the SGU site. Medium-steep fractures make up 31 % and the remaining 20 % are flat-lying fractures. It is possible to distinguish the following sets at the SGU site.

1 N10E;80E
2 N30E;85E
3 N30W;90
4 Sub-horizontal;25

These sets are shown in the semispherical projection in Figure 3, which also includes the sets found in the SAC area.

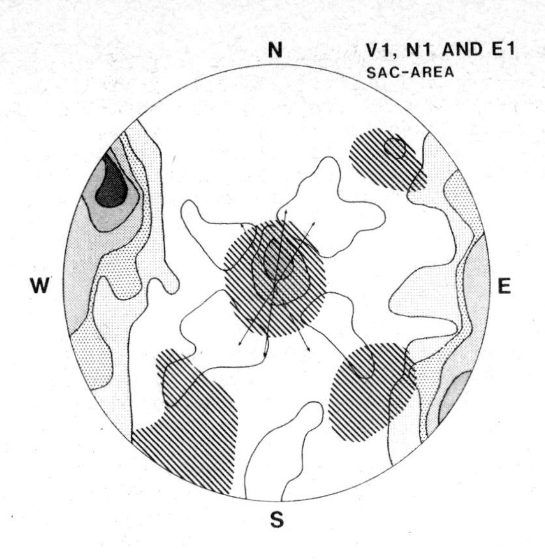

Figure 3. Fracture sets obtained at the SGU site and from the SAC area. Semispherical projection, Schmidt net - lower hemisphere. Contour density 2-4-6-8-10-12%.

The fracture-filling minerals were megascopically classified on the basis of colour, hardness and appearance of carbonates. X-ray diffraction showed that plagioclase mixed with epidote was much more common than expected.

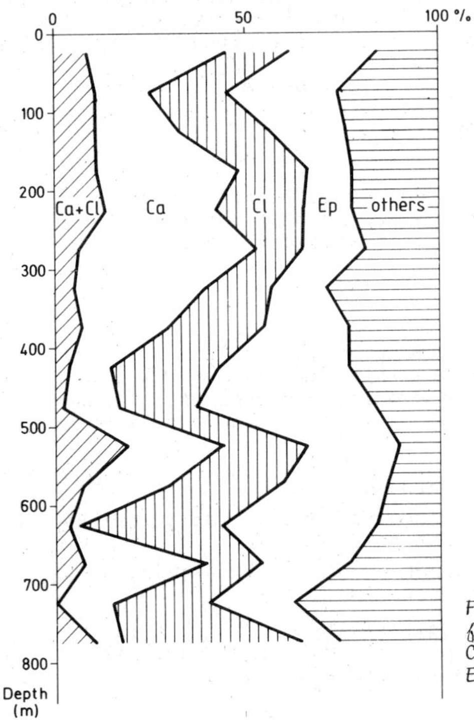

Figure 4. Distribution of fracture filler in V2 versus depth. Ca = Calcite, Cl = Chlorite and Ep = Epidote.

Figure 4 shows the variation of mineral fillers versus depth from borehole V2. Chlorite and chlorite/calcite fillers are roughly constant throughout the full length of the borehole. The most striking change with depth is that calcite shows a marked decrease at a depth of 350 - 450 m with a simultaneous increase in epidote. The group of "other minerals" such as pyrite, flourite, iron oxides, zinc sulphide and clay, comprises a complex group with great variety.

2 RESULTS

2.1 Drilling

Drilling of the horizontal and vertical holes was performed with double core barrels. Boreholes V1, E1 and N1 were drilled with a diameter of 76 mm, V2 with 56 mm. The average drilling rate was 8 - 10 cm/min.

The holes were logged with respect to deviation after drilling. Two methods were used, one based on successive photogrammetric survey (Fotobor), the other based on magnetic properties and orientation in the rock (SGU). The results showed good agreement.

The horizontal boreholes E1 and N1 have a slightly downward inclination in order to avoid entrapment of gas in the hole. The deviation is in the horizontal lane towards the right in both boreholes. In the vertical plane, the deviation for E1 is about 1 m upwards but almost negligible for N1. In general, the deviations obtained are limited and acceptable.

2.2 Core-logging

The four main boreholes in the program were all logged according to a standard procedure, which includes recording of distinctive changes in rock type, colour, grain size and fracture characteristics. These characteristics are listed in tabular form and include the following:

* Depth and type of fracture (open, sealed or induced)
* Orientation with respect to core axis
* Mineralogy of fracture fillers
* Slickensiding

All boreholes penetrate granite to their full length. Dykes and massifs of greenstones and pegmatites are present, mainly as thin veins, but in some sections of the cores with a thickness of several meters.

2.3 Geophysics

Geophysical well-logging plays an important part in the program for determining hydraulic, structural stability and chemical conditions. A standard geophysical well-logging program has been established and carried out in the four main boreholes. This program comprises the following logs:

Type of log	Purpose
Deviation log	Direction and deviation of the borehole
Natural gamma	Rock type, dykes, veins and fracture indications
Single point resistivity	Resistivity of the rock in the borehole wall i.e. conducting minerals and fractures

Resistivity logs	Fracture indications
Temperature logs	In- and outflow zones in the hole
Borehole fluid resistivity	In- and outflow zones in the hole
Spontaneous potential log	Measures anomalies which indicate fracture zones

The results of natural gamma logs presented in Table II show large differences between the holes. The high level in N1 is due to high radon content in combination with the very low water outflow. The radiation level in the granite is most accurately indicated by the values from V1, where the water outflow is high.

Table II. Radiation level in boreholes E1, N1, V1 and V2.

	Average	Peak
V1	65 microR/h	406 microR/h
V2	100 microR/h	250 microR/h
E1	117 microR/h	
N1	250 microR/h	430 microR/h

The temperature of the water in boreholes V1 and V2 is given in Figure 5. The influence of the drift on the temperature can be seen down to about 100 m in V2. From there, the temperature gradient is about 17 degreesC/km down to 480 m, where it decreases to around 15 degreesC/km. From 610 m, the gradient increases to 18 degreesC/km down to the bottom of the hole. The influence of the drift is not evident in V1. Instead, the temperature is higher than in V2 due to the outflowing water emanating from the borehole below 460 m. At the bottom of V1, the temperature is 19.1 degrees C, and at the corresponding level in V2 the temperature is 19.6 degrees C.

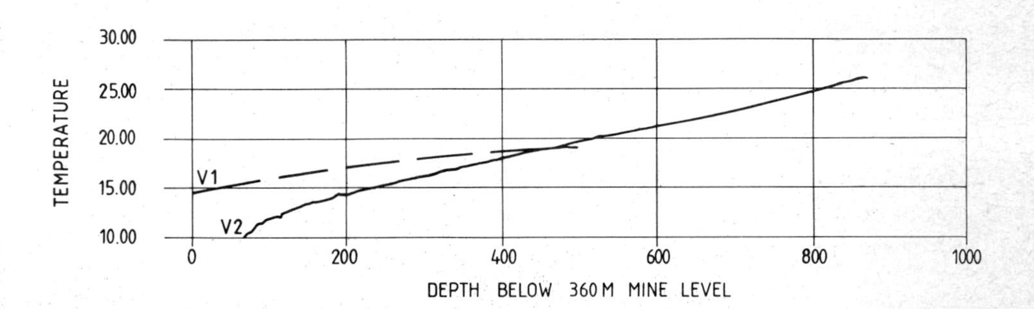

Figure 5. Water temperature versus depth in boreholes V1 and V2.

2.4 Hydrogeology

2.4.1 Hydraulic units

The hydraulic conditions of a crystalline rock mass such as the Stripa granite is characterized by the discontinuities which intersect the rock. The granitic rock matrix is virtually impervious and the main flow paths are made up of the fracture system, zones of fractured or crushed rock and other structural discontinuities.

Most ruptural deformation is concentrated to the superficial part of the rock, which shows a rather high fracture frequency and high hydraulic conductivity. This more fractured part of the rock mass extends down to a depth of about 250 m. Further down the rock becomes more sparsely fractured, with fractures that are sealed to a great extent. The degree of fracturing continues to decrease and reaches its lowest recorded frequency at the 1100 m level.

In the deep-lying rock mass, the water flow seems to be channelled through a few zones of fractured rock. The zone found in the lowermost part of V1 is an extreme example of these flow paths. At these deep levels, it is probable that discrete fracture flow is of minor importance.

The mine itself is one of the most important structures governing the water flow in the area. It acts as a drain, with a drainage threshold that was successively lowered as mining progressed. Measurements were made during the SAC program which illustrated the function of the mine as a draining structure. It can be seen from piezometric recordings /2/ taken at different levels in SBH-1, SBH-2, SBH-3 and DbhV1 that there is a gradient towards the excavations.

2.4.2 Hydraulic conductivity of the rock mass

A great number of hydraulic tests have yielded values on the hydraulic conductivity of the Stripa granite. The large body of data provides a good basis for determinations of the water flow in the granite rock mass around the mine. Table III shows the range in the values from the SAC program and the present program.

Table III. Values of hydraulic conductivity in the Stripa granite obtained from the SAC program and the present program.

Test type	Conductivity range m/s	Ref.
Injection surface holes	5 E-11 - 5 E-8	2
Injection ventilation drift	1 E-12 - 1 E-9	2
Ventilation test	1 E-11	3
Large scale injection	4 E-11	4
V1, fracture zone	7 E-8	
V1, rock mass	5 E-11	
V2, rock mass	1 E-10	
E1	5 E-12 - 4 E-8	

Conductivity is at its maximum in the surface boreholes, about 5 E-8 m/s, while it is 1 E-9 m/s or lower in the tests made down in the mine. The large-scale ventilation and injection tests, both of which are measures of gross conductivity, gave low values, 1 E-11 and 4 E-11 respectively. Those values are probably representative of the rock mass, including minor zones of fractured rock.

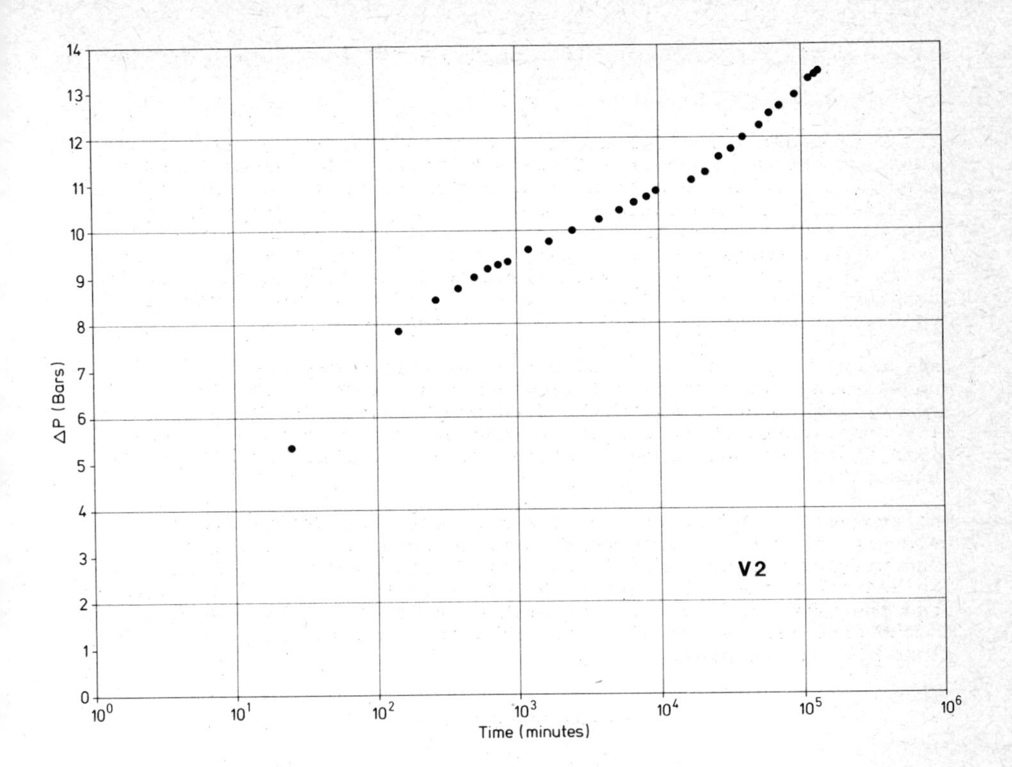

Figure 6. Change in water pressure versus time for pressure build-up test in V2 section 356 - 470.

The hydraulic tests carried out in the present program were all performed as pressure build-up tests, where the natural water flow into the boreholes was used for the build-up. The tests were analysed according to conventional interpretation techniques. An example of a test is shown in Figure 6.

Two zones have been found with relatively high conductivity, one 40 m wide zone in V1 (40 m along the borehole) with a conductivity of 7 E-8 m/s and one 2 m wide zone in E1 with 4 E-8 m/s. Aside from in these zones, the values obtained for conductivity were lower than E-9 m/s.

2.4.3 Hydraulic head

The hydraulic head in rock is determined by geological, hydrometeorological and topographical factors. In the present case it is also dependent to a very high degree on the geometrical configuration of the mine.

The geological factor, which determines the hydraulic conductivity and thus the rate of the groundwater flow in the bedrock, points to a rather low conductivity and consequently a low groundwater flow even at high hydraulic gradients. In the upper part of the bedrock, the groundwater table generally follows the topography, owing to meteorological and geological conditions.

Hydraulic head was measured in boreholes drilled both from the mine and from the surface. Measurements of hydraulic head are normally made in short sections in boreholes tightly sealed off by packers. At the start of such measurements, the head is usually in a transient state and the recording has to be made over a long period of time.

2.4.4 Model calculations

Preliminary and rough calculations were made of groundwater conditions and the groundwater inflow to the mine. The calculations were based on available data on hydraulic conductivity and topographical conditions. The calculations were performed for a veritical plane laid out from the center of lake Rosvalen, through the mine and about 4 km towards NNW. In total, the section was 7 km in length and 2.6 km in depth. The mine was illustrated as two horizontal drifts, each 1000 m in length, at the 410 m and 290 m levels in the mine system. The height of the drifts was taken to be 70 m.

The calculations were carried out using a finite-element program and assuming two-dimensional flow under a steady state. The lower and vertical boundaries to the studied plane were assumed to be no flow boundaries. The groundwater head at the upper boundary was given as the ground surface. At the mine, the head was set at the datum level.

The hydraulic conductivity of the rock mass was assigned different values to illustrate different possible conditions in the rock. The conductivity versus depth distribution used is given in Figure 7. The results of the calculation, given as distance of influence on head and water inflow, are summarized in Table IV. The total inflow to the mine was estimated by assuming the same inflow per m of mine along the entire mine.

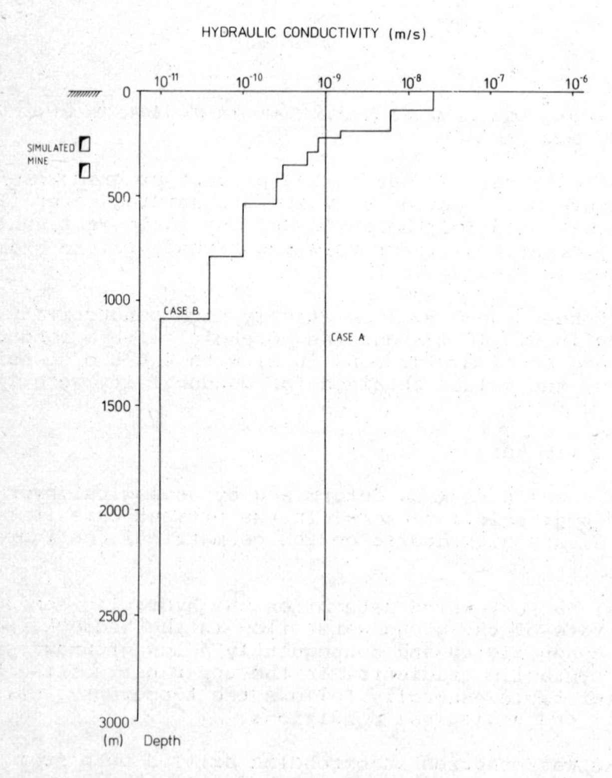

Figure 7. Hydraulic conductivity versus depth used in the model calculations.

Table IV. Results of numerical calculations of the distance of influence and the groundwater inflow to the Stripa mine.

Assumptions made regarding the K-value of the rock mass	Horizontal distance for 50 m influence at the 400 m mine level in km	Groundwater inflow to the mine in l/min
Case A Hom. condition with K=1.E-9 m/s	0.8	73 l/min
Case B Hom.condition with decreasing K-value from 2.E-8 down to 3.E-11 m/s	0.45	96 l/min

The actual inflow to the mine is estimated to be about 600 l/min /5/. The result based on decreasing conductivity values versus depth (Case B) is more reliable, as it is based on actual test results from the area. The groundwater head for this distribution is given in Figure 8, where the impact of the mine is clearly shown.

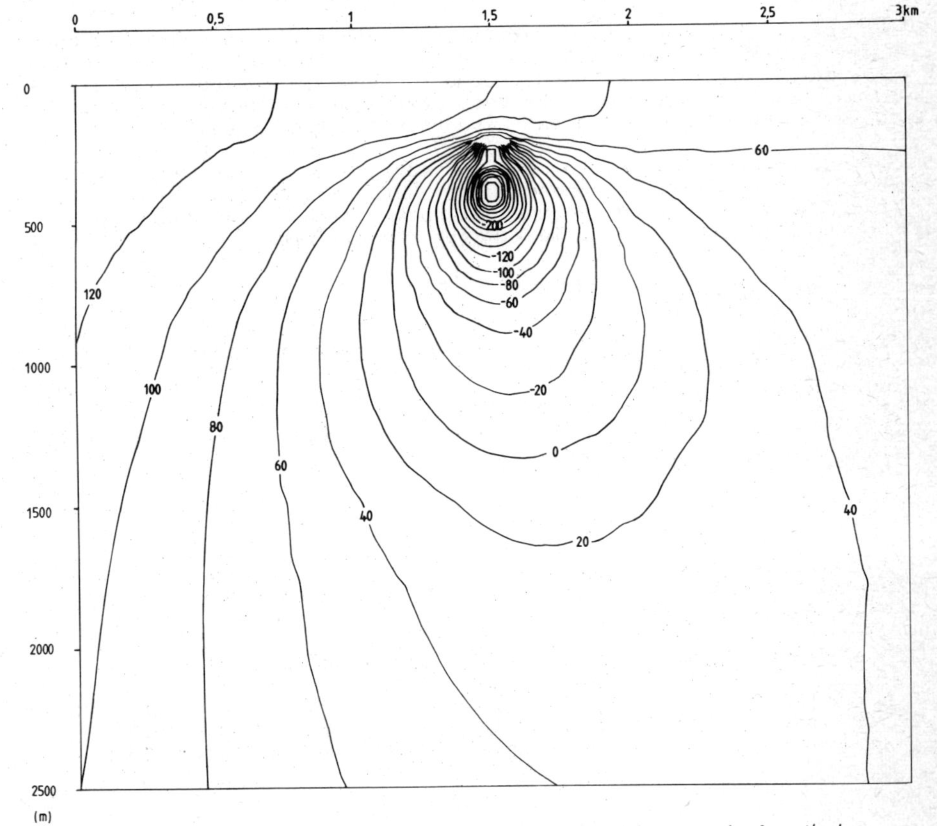

Figure 8. Groundwater head around the mine calculated by numerical method, case B.

2.5 Geochemistry

Priority was given in the program to hydrogeochemical investigations in the vertical boreholes, while only a small chemical program was set up for the horizontal boreholes. Water was sampled for analyses of major, minor and trace elements, along with isotopes and inert gases. The results so far are presented in a separate paper of the Workshop /6/.

3 CONSLUSIONS

The results of the investigations performed have yielded valuable information on instruments and methods used as well as on hydrogeological and hydrochemical conditions at great depth in crystalline rocks. Thus, core-drilling can be carried out with only slight deviations from predicted orientations. Underground hydraulic testing methods should preferrably be carried out by using existing conditions of head and flow instead of injection or withdrawal of water.

Cross-hole testing gives more valuable and accurate information and can be conducted within deep lying areas of interest when performed from underground sites instead of from the surface.

The results obtained show that fracture zones with a hydraulic conductivity of more than 1 E-8 m/s exist even at great depth. The deep rock mass, on the other hand, seems to have low conductivity with no appreciable variation. The composition of the water at great depth is dependent to a great extent on the rock itself.

Further work on the development of underground investigation methods should focus mainly on nondestructive cross-hole techniques aimed at characterizing large volumes of rock, which is necessary for the detailed siting of nuclear waste repositories.

4 REFERENCES

/1/. Carlsson, L. and Olsson, T.: Hydrogeological investigations in boreholes. Summary of defined program. Stripa Project. Technical Report 81-01.

/2/. Gale, J. E., 1982: Hydrogeologic characteristics of the Stripa site. University of Waterloo, Report 003C.

/3/. Witherspoon, P. A., Cook, N. G. and Gale, J. E., 1980: Progress with field inverstigations at Stripa. SAC Report 27.

/4/. Lundström, L. and Stille, H., 1978: Large scale permeability test of the granite in the Stripa mine and thermal conductivity test. SAC report 02.

/5/. Hallén, P. A., 1982: Personal communication.

/6/. Nordstrom, K., 1982: Geochemical characteristics of groundwater at Stripa. NEA/KBS Workshop on in situ Experiments in Granite Associated with Geological Disposal of Radioactive Waste. In press.

APPLICATION OF THE SINUSOIDAL PRESSURE TEST TO THE MEASUREMENT OF HYDRAULIC PARAMETERS

by

J.H. BLACK and J.A. BARKER*

Institute of Geological Sciences, Harwell Laboratory,
Oxfordshire, England.

*Institute of Geological Sciences, Maclean Building,
Wallingford, Oxfordshire, England.

ABSTRACT

The sinusoidal test method is briefly described against the background of oil-industry pulse testing. Although a homogeneous porous medium analysis of the sinusoidal method already exists a new fissured porous medium analysis is derived. The possibility of distinguishing between homogeneous and fissured responses by using a wide range of source frequencies is discussed. The results of a preliminary trial of the method in a granite quarry in Cornwall are given together with a brief outline of the equipment used. Some of the results required the use of the fissured analysis. The directional nature of the results is shown in the form of hydraulic diffusivity vectors between the two boreholes. It is concluded that the method is worth more development.

1. Introduction

The sinusoidal pressure test is a cross-hole technique in which a small zone of one borehole is subjected to a sinusoidal variation of pressure whilst similar zones in adjacent boreholes are monitored. The pressure variation in the source zone is created by a carefully controlled regime of injection and abstraction. The receiver zones should detect a sinusoidally varying pressure which has a smaller amplitude and lags behind the source zone signal. The decrease in amplitude and the retardation (phase lag) of the received signal compared to the source signal depend on a number of the hydraulic properties of the rock between the source and the receiver.

This approach has several advantages and disadvantages when compared to the usual step change aquifer test namely:

Advantages:

1. The diagnostic measurements are phase shift and amplitude, so the start-time for the test is irrelevant, which enables movements of the receiver zones without having to stop the excitation.

2. The testing is essentially point to point with the potential ability to calculate the hydraulic conductivity ellipsoid for each source zone.

3. There is no net discharge so equilibration times are small enabling rapid movements of the source zone position.

4. The oscillating signal is detectable against a changing background pressure (especially important in mines).

5. The frequency of the test can be changed to investigate different components of the bulk-rock hydrogeology, i.e. fracture properties and matrix properties.

Disadvantages:

1. The distance of penetration of measurable pressure fluctuations is less than the step change method.

2. The test requires equipment which is more complicated than straightforward abstraction testing.

3. The property measured is essentially hydraulic diffusivity, a combined function of hydraulic conductivity and specific storage rather than the two separately.

A form of the method (known as "pulse-testing") is already quite widely used in the petroleum industry where the advantage of being able to discern the signal against a varying background pressure due to producing wells is of great economic benefit. Pulse testing differs, however, in comprising alternate periods of abstraction and shut-in producing a "pseudo-square wave" source and a "pseudo-sinusoidal wave" received signal all superimposed on a declining pressure level [1,2,3].

Against this background it was of interest to evaluate whether or not a sinusoidal signal would be detectable over significant distances in rocks such as those proposed for radioactive waste disposal. The theoretical appraisal [4] showed that whilst in fractured crystalline rocks a signal might be measurable up to distances around 100 m, argillaceous rocks would attenuate the signal too rapidly. A point source in a homogeneous rock was chosen for the evaluation since this yields maximum attenuation. Since that initial theoretical appraisal of the technique, the analysis has been developed to cover a number of configurations and a field trial has been performed.

2. Analysis

2.1 Previous Work

The appraisal [4] contained two basis configurations: that of a point source and that of a line source within homogeneous isotropic porous media. Clearly these are not going to be adequate for analysing actual tests in fractured porous media such as granite although tests in single fractures or fracture zones may be sufficiently well modelled by the line source analysis at small scales. A programme of analysis development has therefore been undertaken to cover a number of situations such as a line source with interacting slabs of porous rock, a point source in anisotropic homogeneous rock and a point source with interacting blocks of porous rock. The derivation of the former analysis is presented below since it introduces an interesting facet of sinusoidal testing: the possibility of changing the frequency of the signal in order to investigate either the diffusivity of the fracture system only or of the whole system (fractures and matrix).

2.2 Derivation of an Analysis for a Line Source in Fractured Rock

2.2.1 Model Description

For a sinusoidal test in a fissured formation the following simple model is considered. A periodic vertical line source intersects a series of equally-spaced, horizontal fissures which divide the rock mass into slabs of thickness 2a. It is assumed that the line source interacts only with the fissure system, in which flow is horizontal. Flow in the rock matrix is only vertical and Darcy's law applies throughout the system.

As a result of the vertical symmetry of the fissures, only a single fissure and its adjacent rock matrix is considered. The piezometric heads in the fissure and the matrix, at time t, are denoted $h_f(r,t)$ and $h_m(r,z,t)$ where r is the distance from the line source, and z is the (vertical) distance to the nearest fissure.

2.2.2 Flow Equations

Combining Darcy's law with the conservation equation, flow in a fissure is described by:

$$S\frac{\partial h_f}{\partial t} = \frac{T}{r}\frac{\partial}{\partial r}\left(r\frac{\partial h_f}{\partial r}\right) + K\frac{\partial h_m}{\partial z}(r,0,t) \qquad (1)$$

where T is the transmissivity of a single fissure
S is the storage coefficient of a fissure
and K is the (vertical) hydraulic conductivity of the rock matrix

The equation describing flow in the rock matrix is:

$$S_s\frac{\partial h_m}{\partial t} = K\frac{\partial^2 h_m}{\partial z^2} \qquad (2)$$

where S_s is the specific storage of the rock matrix.

It is convenient to adopt a complex representation of the line source strength so that, if the amplitude in a single fissure is Q_o and the angular frequency is ω, the boundary condition at r = 0 is given by:

$$\lim_{r\to 0}\left(r\frac{\partial h_f}{\partial r}\right) = -\frac{Q_0\, e^{i\omega t}}{2\pi T} \qquad (3)$$

If heads are measured relative to the piezometric surface when there is no pumping, the boundary condition at large distances from the source is represented by:

$$\lim_{r \to \infty} h_f \ (r,t) = 0 \qquad (4)$$

At the fissure/matrix interface, z = 0, the heads will be equal, so:

$$h_m \ (r,0,t) = h_f \ (r,t) \qquad (5)$$

As a result of symmetry, there will be no flow across the centre of a slab of rock matrix, z = a, so the final boundary condition is:

$$\frac{\partial h_m}{\partial z} \ (r,a,t) = 0 \qquad (6)$$

2.2.3 Solution

The solution to equations (1) and (2) with boundary conditions (3) to (6) has the form:

$$\frac{2\pi T}{Q_0} h_f = A \ e^{i(\omega t - \phi)} \qquad (7)$$

where A (Ω,α,β) is dimensionless amplitude and ϕ (Ω,α,β) the phase shift. These can be calculated from:

$$A \ e^{i\phi} = K_0 \ (Z) \qquad (8)$$

where K_0 is a modified Bessel function with complex argument, Z, given by:

$$Z^2 = \Omega \left[i + \left(\frac{i \ \alpha}{\Omega}\right)^{\frac{1}{2}} \tanh \left(\frac{i\Omega}{\beta}\right)^{\frac{1}{2}} \right] \qquad (9)$$

where Ω is the dimensionless angular frequency:

$$\Omega = \frac{\omega \ r^2 \ S}{T} \qquad (10)$$

and the dimensionless parameters α and β are given by

$$\alpha = \frac{4 \ r^2 \ S_s \ K}{S \ T} \qquad (11)$$

and

$$\beta = \frac{r^2 \ S \ K}{a^2 \ S_s \ T} \qquad (12)$$

2.2.4 Asymptotic Behaviour

The variation in behaviour with respect to the three parameters Ω, α and β (Figure 1) is expressed in equation (9), and a few generalisations are possible.

When $\Omega \ll \beta$, equation (9) can be approximated by

$$Z^2 = i \ \Omega \ \left[1 + \left(\frac{\alpha}{\beta}\right)^{\frac{1}{2}}\right] \qquad (13)$$

$$= \frac{i \ \omega \ r^2 \ (S + 2 \ a \ S_s)}{T} \qquad (14)$$

Equation (14) indicates that the formation behaves as a homogeneous medium with a storage coefficient given by the combined storages in the fissures and the rock matrix. This is seen in Figure 1 as the straight line portions on the left hand side where Ω is small. In physical terms the period of each cycle is long enough to allow the fissures and matrix to come to an equilibrium pressure during each cycle. However, as frequency (Ω) increases, the equilibrium behaviour no longer applies and amplitude and phase shift depend increasingly on the fissure properties alone. Thus given a range of test frequencies, it should be possible to determine whether or not it is responding as a

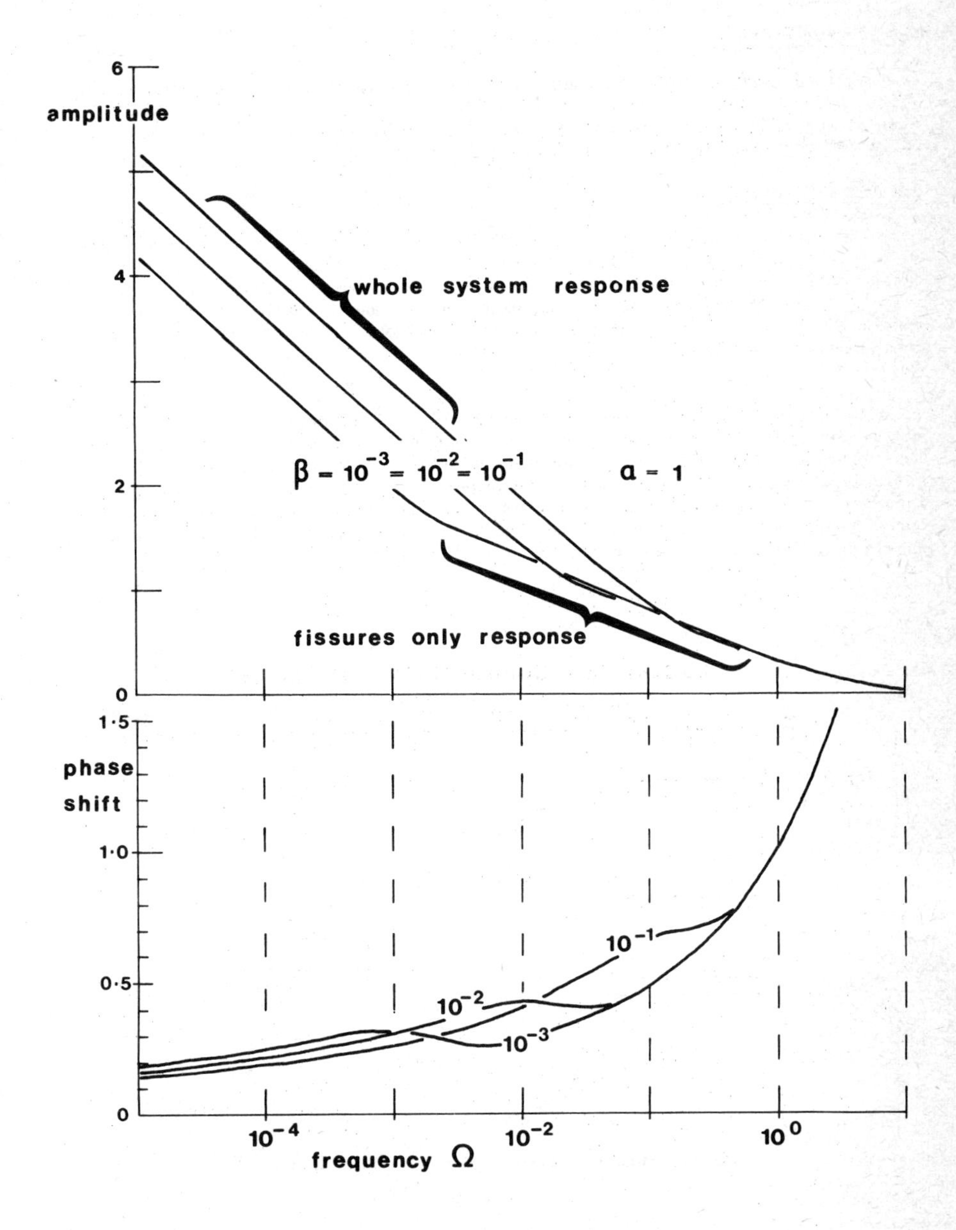

Fig. 1 The dependence of amplitude and phase shift in a dual porosity medium on the parameters of Ω and β (α held constant)

homogeneous or a fissured porous medium. Additionally it can be seen in Figure 1 that the results break away from homogeneous behaviour when $\Omega = \beta$. Thus if fissure and matrix properties remain otherwise unchanged the effect of increasing slab thickness is to decrease β which results in amplitudes being reduced and transition to fissure response occurring at a lower frequency.

3. Preliminary Field Test

3.1 Introduction

A field test of both the theory and the experimental equipment was carried out in December 1981 using two adjacent boreholes in a quarry in Cornwall. Time and equipment were limited in what was a basic experiment, the results of which are reported by Black and Holmes [5].

3.2 Site Description

The quarry is situated on the north-western flanks of a granite intrusion in Cornwall and comprises a long (60 m) narrow (~10 m) cut into a westerly facing hillside. It contains about 30 boreholes of various depths and diameters mostly drilled for heating experiments, and a nearby mine approaches within about 15 m of the quarry floor. The rock in the quarry is a relatively fresh, massive coarse-grained two mica granite [6] with dykes cropping out in the walls.

Two boreholes, with a surface separation of 25 m, depths of 145 and 166 m and variable diameters between 108 and 125 mm were chosen for the testing. They were air flush drilled by down-the-hole-hammer, which resulted in the variable diameter and rough walls. In order therefore to gain downhole information on the fractures a seisviewer was run in an adjacent water flush cored borehole. In the 170 m of borehole, 80 fractures were discerned giving an average frequency of about 1 every 2 m of borehole.

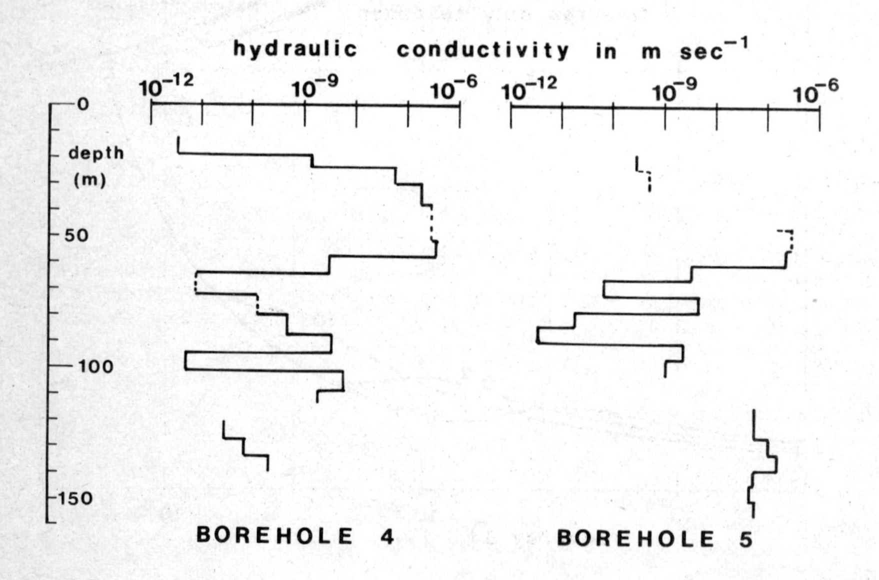

Fig. 2 The variation of hydraulic conductivity with depth in the two selected boreholes

3.3 Borehole Hydraulic Properties

Hydraulic conductivity and environmental pressure were measured in the two selected boreholes using the IGS straddle packer system [7] and a combination of open and closed-tube slug testing. The variation of hydraulic conductivity with depth in the two boreholes is shown in Figure 2. The testing was carried out with a 7 m straddle and measurements were not obtained where the results indicated that testing was significantly affected by leakage past the packers. It can be seen that there are high hydraulic conductivity zones around the 50 m depth in both boreholes and towards the base of Borehole 5. The high hydraulic conductivity zone around 50 m in Borehole 5 was therefore chosen as a source zone since it would enable a transmitted signal to be detected at considerable distance.

3.4 Sinusoidal Pressure Test

3.4.1 Method

Two straddle packer systems were used, one in each borehole, with pressure-transducers measuring the water pressure in both the packered-off zone and the rest of the borehole below the bottom packer. In common with the source zone between 44 and 51 m in Borehole 5 a 7 m straddle was used in the receiver borehole. The equipment used in the source borehole was similar to that used for slug testing but with the addition of a hydraulic pump. This was an oil industry rod-sucker pump coupled to a hydraulic cylinder which is powered in both directions from the surface using a series of hoses in the form of an umbilical.

The hydraulic pump generated the abstraction portion of the pressure wave whilst a surface pump, injecting water via a separate hose, was used to supply the positive part of the wave. Sinusoidal waves were generated by manipulating the controls of the two pumps with the outputs from the source zone and the receiver zone being placed on the same two pen recorders. As can be seen in Figure 3 the rod-sucker pump tended to produce pressure pulses in the source zone though these could be averaged to produce an acceptable abstraction curve. Since the source zone had a high hydraulic conductivity, maximum pressures reached were ±3 m of water head which was compatible with the relatively shallow depth of the source zone below the water table.

In the receiver zone the minimum sensitivity of the equipment was equivalent to a head change of less than 1 mm of water. Given the various testing and drilling operations going on at the quarry within 25 m of the borehole a minimum sensitivity of about 2 mm is more realistic. This is in relation to the signal frequency of 0.008 radians sec^{-1} (1 cycle per 13.2 mins) which was used for all the tests.

At the start of each test, the straddle packers isolating the receiver zone were inflated to an overpressure of about 400 m of water head (600 psi). Slug and pulse tests were then performed in each receiver zone to determine both the zones hydraulic conductivity and its interval response time [8]. The interval response time was required in order to allow for any attenuation or phase lag introduced by the receiver zone characteristics. Once these were complete and a comparatively steady pressure had been reached in the receiver zone, the test was carried out for about four cycles.

3.4.2 Results

Positive results were obtained from 7 receiver zones of Borehole 4 as a result of sinusoidal pressure fluctuations in the 44 - 51 m zone of Borehole 5. The results were analysed in terms of peak attenuation and phase shift and are given in Table 1.

There was some leakage around the packers in the source borehole so that a sinusoidal fluctuation equal in amplitude to about one tenth of the source zone signal was measured in the rest of the borehole below the bottom packer. Measurements in the receiver borehole ceased at a depth of 60 m because the transducer below the bottom packer was not recording any fluctuation and therefore only very small signals could have been expected.

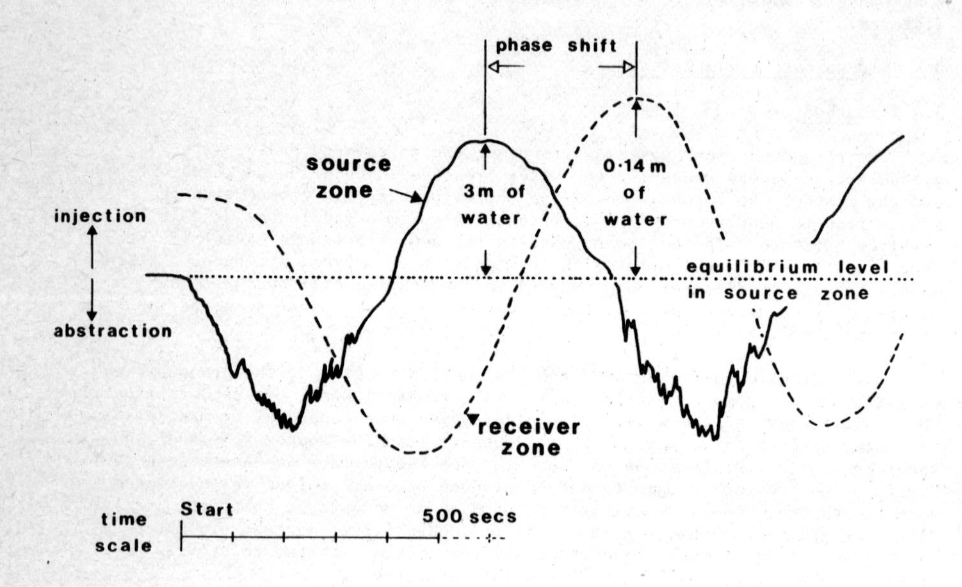

Fig. 3 Example of source and receiver pressures of the start of a test

Table 1

Summary of Results

Depth of receiver zone in Borehole 4 (m below ct)	Interval response time (secs)	Average		Direct distance from source to receiver zone (m)
		Received amplitude*	Phase shift (radians)	
15-22	32	1.48×10^{-3}	0.56	37.9
23-30	3	3.62×10^{-3}	0.41	32.2
30.37	3	3.80×10^{-2}	0.29	28.1
34-41	1.5	4.10×10^{-2}	0.30	26.4
41-48	2.5	8.65×10^{-3}	0.40	24.6
46-53	3	7.54×10^{-3}	0.39	24.5
53.60	10	6.88×10^{-3}	0.33	26.0

$$* \text{ received amplitude} = \frac{\text{maximum amplitude in receiver zone}}{\text{maximum amplitude in source zone}}$$

3.4.3 Analysis

The analysis of such a limited number of results (single frequency, one source zone) requires some idea of the system being tested. For instance, should the rock between source and receiver be considered as a homogeneous porous medium or as a fissured porous medium and does cylindrical or spherical pressure transmission/flow take place? The results were examined, and given the combination of moderate amplitude and phase shift, spherical flow was ruled out. Of the 7 results two conformed entirely to a model involving cylindrical flow into a homogeneous porous medium (Zones 34-41 and 30-37). The other five results were obtainable by considering cylindrical flow into a fissured porous medium broken up into slabs between 20 and 40 mm thickness.

Without more work it would be unwise to overinterpret these initial results but they have been plotted in the form of directional diffusivity in the plane of the two boreholes in Figure 4.

It is interesting to note that, although the high hydraulic conductivity measured in borehole 4 extends right through the zone 30 m to 55 m below casing top, the primary diffusivity is towards the top of this zone. Also the measured hydraulic diffusivity is approximately equivalent to dividing the single borehole hydraulic conductivity results by the whole rock specific storage.

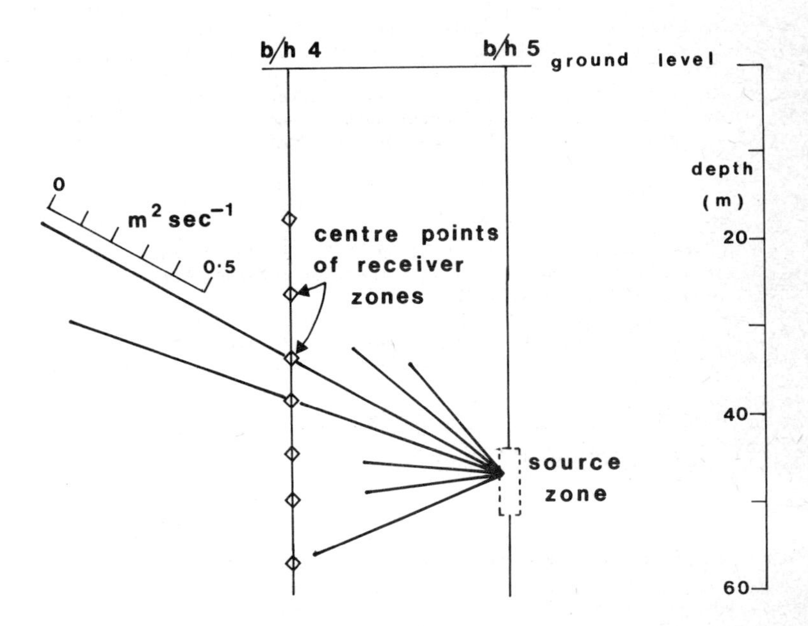

Fig. 4 Sinusoidal results plotted in terms of directional hydraulic diffusivity (length of the line is proportional to the magnitude of the diffusivity)

4. Conclusions

A preliminary trial of the sinusoidally varying pressure method has been carried out in crystalline rock. It has been shown that it is possible to produce an appropriate source signal within a borehole using equipment which is already available. Given the properties of the rock the signal was easily measurable over a 45 m distance using comparatively cheap pressure transducers.

The results of the trial resulted in an expansion of the available analyses to include a variety of conceptual models. These now include analyses for point and line sources, anisotropy and various forms of dual porosity model (slabs, spheres and lumps). One in particular, a line source with slabs of interacting porous rock, is presented and shows the interpretative possibilities of using a wide range of frequencies.

The test also revealed some of its possible advantages over the conventional step change method namely:

1. It was possible to perceive the received wave-form against a changing background.

2. The received wave was seen to diminish to zero within two cycles of ceasing the source function.

3. The results vary with direction.

It is concluded that the approach shows sufficient advantages over conventional tests to justify further development.

5. Acknowledgements

The authors are indebted to Mr. P. Bourke and Mr. B. Watkins who allowed access to the site and who provided useful background information. Most of the field work was carried out by D.C. Holmes and P.J. Smith with additional seisviewer and caliper information from S.L. Shedlock. The work was carried out partly under contract to the European Atomic Energy Community within its research and development programme on waste radioactive management and storage and partly under contract to the U.K. Department of the Environment. This work is published by permission of the Director of the Institute of Geological Sciences (NERC).

References

1. Johnson, C.R., Greenkorn, R.A. and Woods, E.G., Pulse testing: A new method for describing reservoir flow properties between wells, J. Pet. Technol. Trans. AIME, 18, 1599-1604, 1966.

2. McKinley, R.M., Vela, S. and Carlton, L.A. A field application of pulse testing for detailed reservoir description, J. Pet. Technol. Trans. AIME, 20, 313-321, 1968.

3. Pierce, A.E., Vela, S. and K.T. Koonce, Determination of the compass orientation and length of hydraulic fractures by pulse testing. J. Pet. Technol. Trans. AIME, 27, 1433-1438, 1975.

4. Black, J.H. and Kipp, K.L. Jr. 1981. Determination of hydrogeological parameters using sinusoidal pressure tests: a theoretical appraisal. Water Resources Res. 17: pp. 686-692.

5. Black, J.H. and Holmes, D.C. Hydraulic testing in granite using the sinusoidal wave method. Rep. Inst. Geol. Sci. ENPU-82-5.

6. Bourke, P.J., Bromley, A., Rae, J. and Sincock, K. 1981. A multi-packer technique for investigating resistance to flow through fractured rock and illustrative results. In Proc. of OECD/NEA Workshop "Siting of radioactive waste repositories in geological formations". Paris, March 1981, pp. 19-26.

7. Holmes, D.C. 1981. Hydraulic testing of deep boreholes at Altnabreac: Development of the testing system and initial results. Rep. Inst. Geol. Sci. ENPU-81-4.

8. Black, J.H. 1978. The use of the slug test in groundwater investigations. Water Services 82: 174-178.

Session 3

GEOCHEMICAL AND MIGRATION INVESTIGATIONS

Chairman - Président

O. HEINONEN

(Finland)

Séance 3

RECHERCHES DANS LE DOMAINE GEOCHIMIQUE ET DE LA MIGRATION

ISOTOPE HYDROLOGY AT THE STRIPA TEST SITE

P. Fritz, J.F. Barker, J. Gale*
University of Waterloo
Waterloo, Canada

ABSTRACT

Groundwater occurrences at the Stripa test site were investigated with geochemical and isotopic techniques. The results show that it is possible to identify and characterize individual groundwater systems. Special attention was given to the occurrence of deep groundwaters which discharge from fractures intersected by boreholes at depths in excess of 800 m. These waters have elevated salinities which are possibly related to the admixture of minor amounts of fossil sea water. This interpretation would agree with observations made at numerous sites in Sweden although it requires for the Stripa groundwaters very extensive rock-water interactions.

Environmental isotope analyses on water samples, aqueous species and facture minerals do not contradict this explanation although other sources for the salt cannot be totally excluded. However, the isotope results show conclusively that the waters are old groundwaters which formed in an environment which differs markedly from the modern environment at Stripa and isotope distributions reflect the geochemical history of these waters.

RESUME

Les eaux souterraines du site expérimental de Stripa ont été étudiées à l'aide de méthodes géochimiques et isotopiques. Les résultats indiquent que l'on peut identifier et décrire des systèmes individuels d'eaux souterraines. Une attention spéciale est accordée aux eaux souterraines profondes qui se dégagent des fractures croisées par des forages à des profondeurs dépassant 800 m. Ces eaux sont caractérisées par une haute teneur en sel, ceci probablement en raison des mélanges avec des petites quantités d'eaux de mer fossiles. Cette interprétation s'accorderait avec des observations relevées en plusieurs endroits en Suède, quoique pour les eaux souterraines de Stripa ceci requiérerait une interaction roche-eau très étendue.

Les analyses de l'environnement des isotopes faites sur des échantillons d'eau, des espèces ioniques et des minéraux de fracture ne contredisent pas cette explication, bien que l'on ne puisse exclure totalement d'autres sources pour le sel. Néanmoins, les résultats isotopiques démontrent que les eaux sont des eaux souterraines anciennes qui ont été formées dans un environnement très différent de l'environnement actuel à Stripa, et les distributions d'isotopes reflètent l'évolution géochimique de ces eaux.

* Present address : Memorial University of Newfoundland.

INTRODUCTION

A first assessment of the geochemistry and isotopic characteristics of the groundwaters at the Stripa test site was presented by Fritz et al. [1,2]. In the following additional analytical work was done under contract with the Lawrence Berkeley Laboratories, University of California. Most data accumulated during this phase have not been published but a summary of all earlier observations were substantiated and this presentation discusses the principal conclusions based on chemical and isotopic data obtained on water samples, aqueous species and fracture calcite. Isotope data are emphasized as a detailed discussion of geochemical data will be presented by K. Nordstrom (this volume). Specific aspects of the isotope investigations will also be presented in forthcoming publications in which our data will be compared with results obtained from follow-up studies.

GEOCHEMISTRY

The most remarkable feature of the groundwater chemistry at the Stripa test site is the occurrence of deep groundwaters with dissolved solids contents that approach 1000 mg/L. They are Na-Ca-Cl waters with very low Mg and K concentrations and low alkalinities. Table I summarizes the water geochemistry measured during the 1979 sampling period and the significant differences in chemical composition and total dissolved solids between shallow groundwaters and waters discharging at various mine levels.

Table I. Geochemistry of Stripa groundwaters

Sample no.		21-29	18-1	85-15	42-15	59-3
borehole		PW 5	PW 1	SBH 3	M3	V2
depth (m)		21	80	89-104	330-344	810-838
date		79-5-15	77-9-27	79-5-27	78-11-17	78-11-20
pH		6.2	7.8	7.9	9.1	9.6
Ca	mg/L	5.7	37	33.9	15.4	110
Mg	mg/L	1.5	12	4.5	.26	.18
Na	mg/L	2.7	24	12.5	53	177
K	mg/L	1.1	1.1	1.7	.2	.55
Cl	mg/L	23	16	3.7	56	449
SO_4	mg/L	13.1	13	8.8	2.5	55.4
HCO_3	mg/L	13	206	144	75	7.8
SIO_2	mg/L	10	8.6	12.5	11.1	10.2

The chemical characteristics of the different waters have remained virtually constant since sampling began in 1977. For example, borehole M 3 in the timescale room has been sampled over a four year period during which it discharged in excess of 300,000 liters of water. As shown in figure 1, only minor temporal variability exists and the variations seen during the 1978 season are probably due to interferences from hydrological testing.

The chemical compositions of the deep groundwaters display a similar constancy which documents that the differences between deep and shallow groundwaters are not transient phenomena due to contamination with drill fluids or disturbances caused by the mining/excavation activities. The interpretation of these data must, therefore, be done in the context of geological and hydrological parameters and must include regional as well as local considerations. This is particularly important for discussions about the origin of the more saline waters below the 800 m level.

The occurrence of saline groundwaters in the crystalline rocks of the Fennoscandian Shield and overlying sedimentary sequences is not uncommon. They are not only found at great depth but are also encountered in the search for shallow, potable water supplies [4,5].

Since the last glaciation large parts of Southern and Central Sweden have been invaded by saline waters at least twice. The extent of the inundation is indicated by the Holocene marine limits shown in figure 2. Between about 10,000 and 9,000 a B.P. the brackish Yoldia Sea covered an extensive area, and between 7,000 and 3,000 a B.P. the Litorina Sea was present west and south-west of Stockholm. During both transgressions existing groundwaters would have been replaced by and mixed with the saline "surface" waters and, as a result, "a map of Sweden depicting recorded occurrences of boreholes yielding groundwater in which the chloride content exceeds 300 mg/L coincides very well with a map of areas that were inundated by postglacial seas" [4].

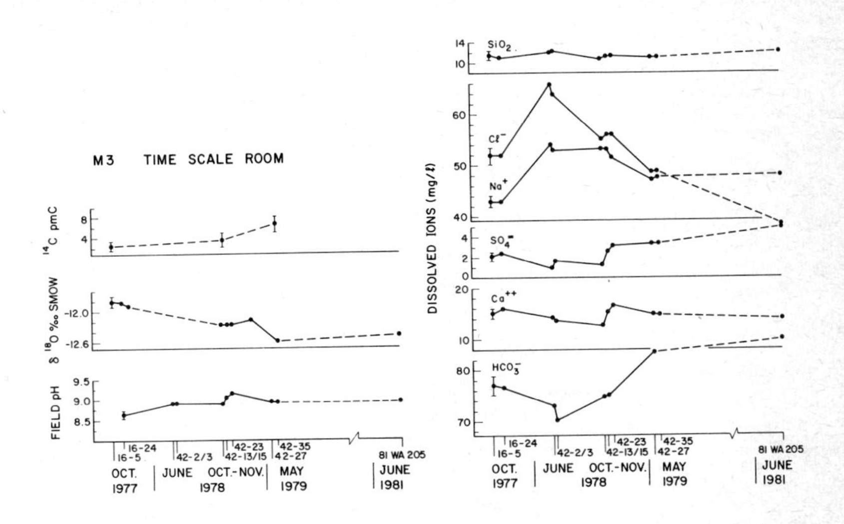

Figure 1. Chemical and isotopic variations for borehole M 3 (time scale room) for samples collected between Oct. 1977 and June 1981

The saline waters discharging today in the Stripa excavations could thus have infiltrated during one of these events, the Yoldia invasion being the more likely one. The occurrence of saline groundwater in the Stripa area is, therefore, not too surprising, but astonishing is the chemical composition of the deep groundwaters which do not reflect a simple mixture of fresh groundwater with marine waters. Such mixing accounts for many of the saline groundwaters encountered elsewhere in this Shield and which are characterized by Ca/Mg ratios greater than 3, whereas in the Stripa waters this ratio is above 100 [2].

If the deep groundwaters at Stripa have a marine component then one would have to accept that these waters were modified through rock-water interactions. Table II summarizes the expected chemistries if local groundwater were mixed with water with marine composition. The calculations assume, that chloride is a conservative ion.

The data in table II show, that the most dramatic changes occur in the Ca, Mg, K and HCO_3 concentrations: Considerably more calcium is present than can be explained by mixing of local fresh water with sea water and virtually all Mg and K are lost. For Na the loss is somewhat smaller as more than 70% is still present. Sulphate appears to behave rather conservatively. The bicarbonate value in mine waters are all very low and strongly suggest that bicarbonate was lost from these waters through secondary processes such as calcite precipitation [1,2].

The important reactions which could account for the observed changes have been summarized by Jack [7,8] and will be discussed by K. Nordstrom (this volume). However, it is relevant for the interpretation of the isotope data that waters with the chemical characteristics of the Stripa groundwaters do occur elsewhere in central Sweden. Figure 2 indicates four locations (A, B, D, E) which were identified by Enquist [6] and chemical compositions are compared in figure 3. The salinites of these wells are higher than those observed to date in the Stripa samples, but the similarity of chemistries suggests that regional rather than local processes must account for the geochemical/isotopic evolution of these waters.

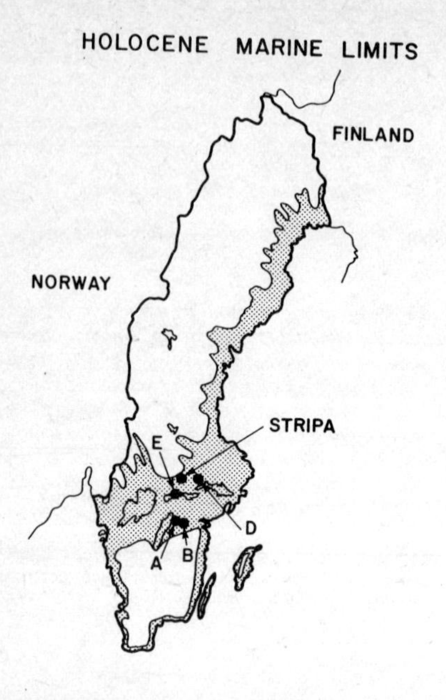

Figure 2. The Holocene marine limits in Sweden [5] and sites with groundwater chemistries similar to those observed at Stripa [6].

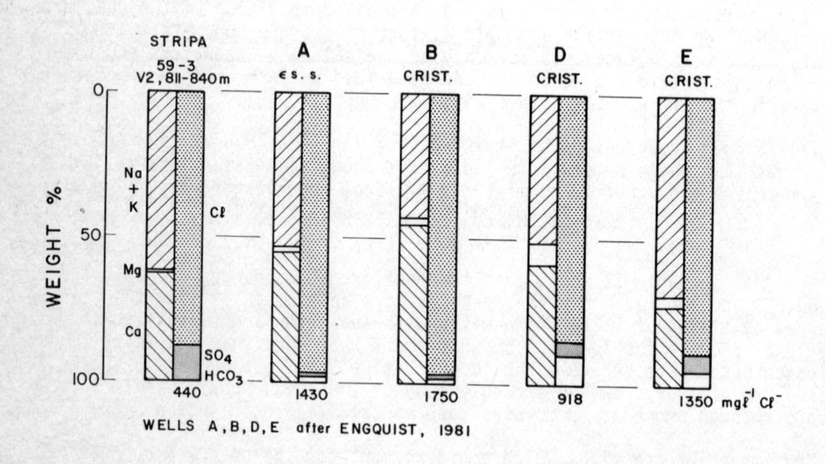

Figure 3. Bar-diagrams depicting chemical characteristics of groundwaters found in Central Sweden and Stripa. Localities are indicated in figure 2.

Table II. Chemistries of groundwaters after mixing with marine water

	sea water mg/L	SBH 3 mg/L	59-3 mg/L	mixed water mg/L
Cl	19350	3.7	449 (620)	
Ca	411	33.9	110 (173)	42.6 (46.0)
Mg	1290	4.5	.18 (.28)	34.3 (45.6)
Na	10760	12.5	177 (266)	261.8 (356.4)
K	399	1.7	.55 (1.4)	10.9 (14.3)
SO_4	2710	8.8	55 (100)	71.5 (95.2)
HCO_3	141	144	7.8 (8.5)	143.9 (144)

ENVIRONMENTAL ISOTOPE ANALYSES

Chemical analyses on the Stripa groundwaters show marked differences between shallow and deep groundwaters. Hydraulic data suggest, that the deep waters will participate in the filling of the excavation and, as a consequence, water movements through the filled excavation might occur. To recognize this is of importance in the assessment of repositories and a knowledge of hydrogeological regimes under predisposal conditions is necessary. Such information cannot be provided through physical hydrology alone but is accessible through environmental isotope analyses. It is thereby understood that isotope data are to be considered in an integrated approach which includes geochemistry and hydrogeology.

During this investigation we have employed a variety of isotope techniques in order to obtain information about the origin of the water and its dissolved load, water ages and rock-water interactions. Specifically the following type of analyses were done:

a) Deuterium (^{2}H) and ^{18}O analyses on water samples were used to characterize/distinguish different groundwaters and to help define their origin and subsurface history.
b) Tritium (^{3}H) and radiocarbon (^{14}C) determinations are the most conventional tools to age date the groundwaters and were especially useful to distinguish modern recharge from older groundwater.
c) ^{13}C and ^{18}O abundances in fracture calcites provide an insight into the origin of the carbon and permit comments on past groundwater regimes in these fractures systems.
d) ^{13}C in the aqueous carbonate species reflects the recharge environments of groundwater or geochemical processes which may take place in the subsurface.
e) ^{18}O and ^{34}S in aqueous sulphate were determined to identify the origin of this ion. ^{18}O in sulphate has a special importance since oxygen isotope exchange between water and sulphate is very slow and can be used as an indirect age indicator.
f) A number of isotopes in the uranium decay series were used to assess geochemical reactions which affect the isotopic abundances of uranium isotopes and daughter products. The suitability of the isotope series for water age determinations was also investigated.
g) Rare gas analyses and rare gas isotopes were used as indicators for groundwater recharge conditions and as indirect age dating tools.

Although all possible tools need not be considered in studies related to the construction of a repository, all contribute to the general understanding of geochemical and hydrogeological systems which surround a repository. For example, Andrews et al. [9] show that one interpretation of the uranium series data could lead to the conclusion that the deep Stripa waters are many tens or even hundreds of thousands of years old. However, the data also indicate that open system conditions prevail and that, therefore the "uranium series method cannot be applied to groundwater age determinations at the Stripa site" Andrews [10]. Despite this, uranium series data show that the deep waters are at least several thousand years old. Furthermore the analyses provide potential information about the mobility of radionuclides of the type to be deposited in a nuclear waste repository.

In this study the rare gas analyses substantiated findings and observations based on ^{18}O and ^{2}H analyses. Helium concentrations, a gas which is tied to the uranium series, shows that active diffusion through the rockmass into moving groundwaters takes place. The recognition of diffusion as active process is very important not only for discussions on radio-isotope migration from a repository but also because it can influence the interpretation of certain environmental isotope data. Theoretical considerations suggest that not only will gas diffuse into the water but it is also possible that dissolved constituents diffuse from open fractures into the pore fluids. The loss can be significant if the isotope is present in very small amounts as is the case of ^{14}C in aqueous carbonate [11,12].

The most direct indications of groundwater origin are obtained from determinations of those isotopes which are part of the water molecule, i.e. deuterium, tritium and ^{18}O. In previous reports we have shown that significant differences in deuterium and ^{18}O contents exist between shallow and deep groundwaters, the distribution of all $\delta^{18}O$ values obtained by us is shown in figure 4. The much lower ^{18}O concentrations in the deep waters are evident. Since the deep waters are free of tritium we concluded that they did not infiltrate recently in the local area. Comparison of deuterium and ^{18}O concentrations show, however, that they have a meteoric origin. Here it must be noted that a 5% contribution of sea water would not be recognizable. Thus these itosope tools do not help to evaluate the possible presence of fossil sea water or any other saline water which might have mixed with normal meteoric waters.

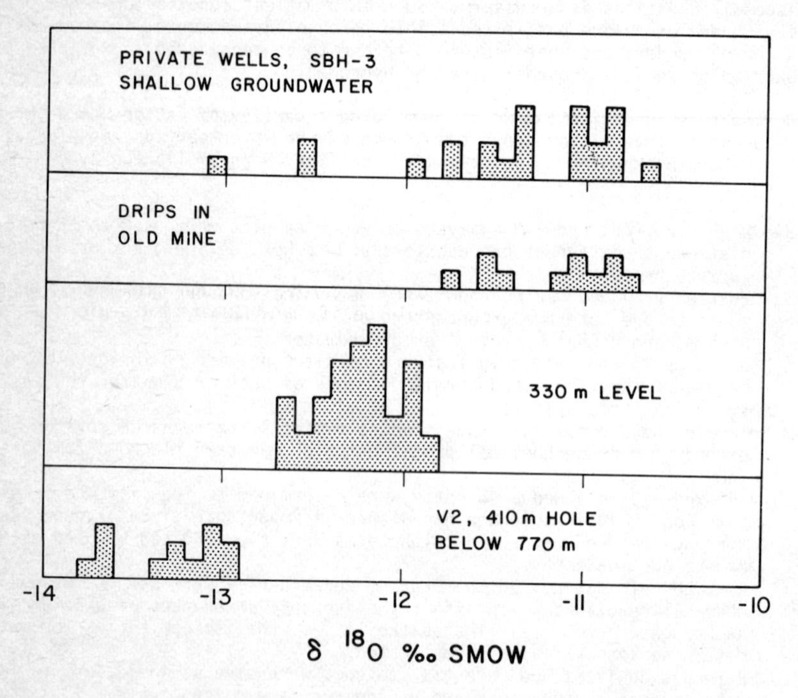

Figure 4. Histograms of $\delta^{18}O$ values in groundwaters at the Stripa test site. Included are data from repeated sampling of the same sampling site.

The most important observation made on the basis of ^{2}H and ^{18}O analyses is that the deep waters, i.e. the fresh water component of the deep waters if they are mixed, could not have formed under the climatic conditions which exist today in the Stripa area. They infiltrated in a much cooler climate and must either have travelled in the subsurface from colder inland mountain areas or formed locally under cooler climatic conditions. In either case we must assume that these waters had a long subsurface history. Their possible subglacial origin was discussed earlier [1,2].

The ^{13}C analyses done on aqueous carbonate and calcites indicate that the waters must have recharged through vegetated terrain, provided that ^{13}C depleted carbon was not generated through oxidative bacterial reactions within the aquifers. Bacterial activities are possible: Calcite samples collected in SBH 2 at a depth of approximately 100 m are strongly enriched in ^{13}C. Their composition is typical for carbonates formed in an environment in which methane producing bacteria are active. At present there are no indications for either bacterial reduction or bacterial oxidation of organic matter in these waters and, therefore, infiltration through soil is the most likely explanation for the observed aqueous carbon compositions. However, two calcites from fracture surfaces encountered in borehole SBH 1 had $\delta^{18}O$ values below -25‰ which suggests that in the past glacial meltwaters migrated through the fracture system and reached depths of at least 300 m.

The presence of biogenic soil(?)-carbon would limit recharge to pre-or post-glacial times. The ^{14}C activities of these waters are very low and correspond to values expected for carbon which is older than 25,000 years. It is virtually impossible to transform these measured radiocarbon activities into water ages but it is noteworthy that ^{14}C analyses done in Finland on saline groundwaters in crystalline rocks gave high activities reflecting water ages below 5,000 years [13]. In this case the chemical data agreed with the interpretation that young groundwaters had mixed with fossil sea water. At Stripa, an interpretation of ^{14}C data yielding young ages would require that either all ^{14}C was lost very quickly through geochemical reactions or diffusive loss into the rock matrix [11] or that sea water mixed with old groundwater which already had very low ^{14}C concentrations. The latter appears to be a strong possibility since invading sea waters will not only displace pre-existing groundwater but mix with it.

The presence of a fossil sea water component in the Stripa groundwaters is not contradicted and possibly supported by the isotopic composition of the aqueous sulphate. Figure 5 presents ^{34}S analyses done during this study and here too a significant difference between shallow and deep groundwaters emerges. There is little doubt that the sulphate in the shallow waters reflects atmospheric input from fossil fuel fallout and, probably, the oxidation of reduced sulphur. The observed isotopic compositions are compatible with this interpretation. Should the sulphate in the deep groundwaters have a similar origin, then its present composition could only be explained if biological or inorganic redox reactions played a major role in the geochemical evolution of these waters. Although this cannot be excluded, the most simple explanation would be if the sulphate had a marine origin with some minor addition of mineral sulphate from the rocks of the area. As shown in figure 5, Quaternary marine sulphate has a sulphur isotopic composition which is close to values measured in the deep waters.

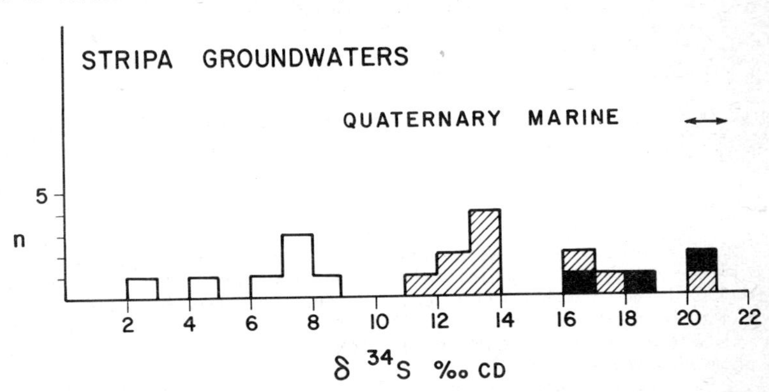

Figure 5. Histograms of $\delta^{34}S$ values in sulphates from Stripa groundwaters. Included are data from repeated sampling at the same site.

The samples collected at the 330 m level boreholes are intermediate in composition which could reflect mixed sources and/or secondary processes. Further work on this aspect is in progress and will be presented at a later date [14].

A small number of aqueous sulphate samples have been analysed for the ^{18}O contents and results are summarized in figure 6. The difference between shallow and deep waters at Stripa is again apparent. Furthermore, the $\delta^{18}O$ values of the deep sulphate are also close to values known for marine sulphates which averages +9.7 ‰. Our values are slightly lower and are close to +8.5 ‰ which may well reflect minor differences between open marine sulphate and the sulphate present in the precursors of the Baltic Sea or cold be due to minor admixtures of isotopically light sulphate as it is generated in the shallow groundwaters.

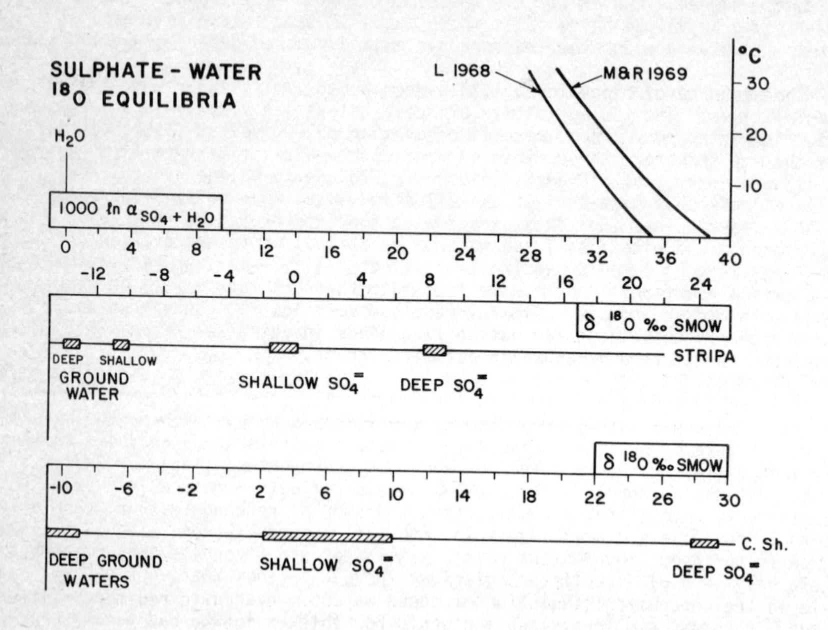

Figure 6. Oxygen-18 in the sulphate-water system. The top of the figure shows the expected ^{18}O fractionation between aqueous sulphate and water [15,16]. The central part gives the range of $\delta^{18}O$ values measured in the Stripa samples and the bottom part shows data from the Canadian Shield [17]. Isotopic equilibrium has been achieved in the latter.

Another important observation is that the deep sulphate is not in isotopic equilibrium with the water. The equilibrium difference in ^{18}O would be close to 30 ‰ (see top of figure 6) and under equilibrium conditions the expected value would be + 17.1 ‰, a value quite different from the measured one. This isotope exchange is strongly pH dependent, and the estimated halftime of reaction is close to 150,000 years in the deep Stripa groundwaters which have pH values close to 10 but only 5,000 years if the pH were about 7 [16]. Yet even an approximate age estimate is difficult to make. The initial $\delta^{18}O$ value of the sulphate is unknown as is the degree of mixing of multiple sources. Furthermore, the pH of the deep water might not have been near 10 throughout its total residence time and ^{18}O exchange between sulphate and water thus would have been much faster than under today's condition. However, should the sulphate have a marine origin and should the pH have been high during most of the subsurface history of these waters, then the lack of isotopic equilibrium would indicate an age of at most a few thousand years. If these conditions are not met, then a preglacial marine origin cannot be excluded, although it does appear very unlikely that these groundwaters were in the subsurface for hundred thousands of years as originally suspected on the basis of uranium series dating [1].

CONCLUSIONS

The preceeding discussions demonstrate that it is possible to identify and characterize different groundwater systems in the granite of Stripa on the basis of their isotopic and geochemical compositions and to show that the deep groundwater system is considerably older than the near surface waters. These deep groundwaters have a very complex history and it is difficult or even impossible to quantify all sources and processes which participated in their genesis. Despite this some general remarks are possible.

First, one must again address the concept of "water age". Throughout this discussion we mentioned "ages" and it might appear that the term is used to describe the age of a water mass. This is not so. For example, we show that it is very likely that a significant amount of mixing has occurred between different waters. Therefore, the term "water age" can only refer to estimates of approximate residence times of the bulk water sample. At Stripa and in Central Sweden it is possible, even likely, that very old groundwater with low salinity was mixed with post-glacial sea water and subsequent further mixing could have occurred with much more recent fresh water. The wide range of salinities observed in the groundwaters of Central Sweden would support this assumption as a regional concept. Thus, the determination of absolute water ages which refer to a single water mass is not possible for the groundwaters in the Stripa granite.

In the early reports we emphasized that available data did not provide a unique answer about the origin of the elevated salinities observed in the deep waters. However, evidence appears to accumulate which shows that regional processes dominate over granite specific reactions in the geochemical evolution of these waters. These include the possible mixing of fresh water with marine waters as well as geochemical reactions such as ion exchange, the dissolution of feldspars, precipitation of carbonate and neoformation of clay minerals. Such reactions could explain the loss of Mg, K and HCO_3 as well as the gain of Ca. This interpretation is also compatible with most of the isotope data although minor "adjustment" through less important secondary processes is necessary for some isotope abundances.

Problems of interpretation will arise, if one accepts the waters which discharge at the 300 m level as being simple intermediaries in a geochemical evolution. They may represent this to some degree but it is likely that these waters have seen more mixing after the retreat of the Litorina Sea than the deep waters. Fracture mineral analyses show that at least to a depth of 300 m hydraulic regimes readily respond to changes in surface conditions. For this reason we have not discussed these intermediate systems in detail although they warrant considerable attention. Analyses must include, however, the facture minerals since they provide a major basis for discussions of events which preceed the present groundwater systems.

Considerations of the isotope data at the Stripa test site reveal the true complexity of the geochemical evolution of the groundwaters which discharge from boreholes drilled from different mine levels. However, in conjunction with the geochemical data they permit a clear distinction of different flow systems and thus contribute directly to the hydrogeological programme. Future work should attempt to take advantage of this knowledge as it could permit the development of mathematical models with defined boundary conditions which could not be recognized from a physical analysis of the hydraulic regimes at the Stripa site.

ACKNOWLEDGEMENTS

Our special thanks go to Dr. P. Witherspoon (U. of California) who supported this programme for several years and whose enthusiasm and interest was an important element in the data acquisition and interpretation. Numerous discussions with colleagues at the SGU and KBS in Sweden, the IAEA in Vienna, Dr. J. Andrews, U. of Bath, Dr. J. Fontes, U. of Paris, and researchers at the USGS influenced this study and their input is gratefully acknowledged.

The programme was supported by funds from contract #W-7405-ENG-48 to the Lawrence Berkeley Laboratories under P.O. 4783902, through WRI contract 803-12 and by funds provided through the National Research Council of Canada (Grant A7954).

REFERENCES

[1] Fritz, P., Barker, J.F. and Gale, J.E. : "Geochemistry and Isotope Hydrology of Groundwaters in the Stripa Granite," Univ. of California, Lawrence Berkeley Laboratories, Berkeley, CA, Rep. LBL-8285, pp. 107 (1979).

[2] Fritz, P., Barker, J.F., Gale, J.E., Andrews, J.N., Kay, R.L.F., Lee, D.J., Cowart, J.B., Osbond, J.K., Payne, B.R. and Witherspoon, P.K. : "Geochemical and Isotopic Investigation at the Stripa Test Site (Sweden), "Proc. Symp. Underground Disposal of Radioactive Wastes, IAEA, Otaniemi, Finland, July 1979, Symp. 243/6, 341-365 (1980).

[3] Fritz, P., Barker, J.F. and Gale, J.E. : Summary of Geochemical Activities at the Stripa Test Site during F.Y. 1979/1980. W.R.I. Report #803-12. Mscrpt. (1980).

[4] Nordberg, L. : Problems in Sweden with intruded and fossil groundwater of marine origin. IN: Proc. Seventh Salt Water Intrusion Meeting, Uppsala, Sweden, S.G.U. report 27, 20-23 (1981).

[5] Lindewald, H. : Saline Groundwaters in Sweden. Proc. Seventh Salt Water Intrusion Meeting. Uppsala, Sept. 1981, 24-32 (1981).

[6] Enquist, P. : Some Wells with High Content of Chloride in Central Sweden. IN: Proc. Seventh Salt Water Intrusion Meeting, Uppsala, Sweden, S.G.U., report 27, 33-39 (1981).

[7] Jacks, G. : "Chemistry of Some Groundwaters in Igneous Rocks", Nordic Hydrology 4 (4), 207-236 (1973).

[8] Jacks, G. : Groundwater Chemistry at Depth in Granites and Gneisses. KBS, Rept. 88, pp. 28 (1978).

[9] Andrews, J.N., Giles, I.S., Kay, R.L.F., Lee, D.J., Osmond, J.K., Coward, J.B., Fritz, P., Barker, J.F. and Gale, J.E. : Radio-Elements, Radiogenic Helium and Age Relationships for Groundwaters from the Granites at Stripa, Sweden. Geochim. Cosmochim. Acta 46, 1533-1543 (1982).

[10] Andrews, J.N. : Radio-Elements and Inert Gases in the Stripa Groundwaters. Mscrpt. 1982.

[11] Neretnieks, I. : Age Dating of Groundwater in Fissured Rock. Influence of Water Volume in Micropores. Wat. Resourc. Res. 17, 421-422 (1981).

[12] Sudicky, E.A. and Frind, E.O. : Carbon-14 Dating of Groundwater in Confined Aquifers: Implications of Aquitard Intrusion. Wat. Resourc. Res. 17, 1060-1064 (1981).

[13] Donner, J.J. and Junger, H. : Radiocarbon, Dating of Salt Water found in Wells Drilled into the Bedrock in the Coastal Area of Finland. Bull. Geol. Soc. Finland, 47, 79-81 (1975).

[14] Fontes, J.C. and Fritz, P. : The Isotopic Composition of Sulfate in Groundwaters at the Stripa Site, in prep.

[15] Mizutani, Y. and Rafter, T.A. : Oxygen Isotopic Composition of Sulphates. N.Z. J. Sci. 12, 54-59 (1969).

[16] Lloyd, R.M. : Oxygen Isotope Behavior in the Sulfate-water Systems, J. Geophys. Res. 73, 6099-6110 (1968).

[17] Fritz, P. and Frape, S.K. : Saline Groundwaters in the Canadian Shield, A First Review. Chem. Geol. 36, 179-190 (1982).

PRELIMINARY DATA ON THE GEOCHEMICAL CHARACTERISTICS OF GROUNDWATER AT STRIPA

D. Kirk Nordstrom
U.S. Geological Survey, MS-21
345 Middlefield Road
Menlo Park, CA 94025

ABSTRACT

The deep granitic groundwaters at Stripa are characterized by decreasing alkalinity, increasing pH and increasing salinity (dominantly Na, Ca, Cl and SO_4) with depth. Furthermore, Ca/Mg, Br/Cl and Ca/Cl weight ratios are anomalously high, whereas Mg/Cl and Na/Ca are anomalously low compared to seawater. New evidence from fluid inclusion leaching of V1 and V2 drillcores suggests that the salinity may be derived from salts or brines contained in the Stripa granite and associated metamorphic rocks (leptite). This evidence implies that continued increases in salinity can be expected at greater depths, especially if the hydraulic conductivity remains low or decreases further.

INTRODUCTION

One of the essential requirements for the safe storage of radioactive wastes in a subsurface geologic environment is knowledge of the groundwater geochemistry. Groundwater geochemistry can provide information on the potential corrodability of waste cannisters, the reactivity and migration potential of leaking radionuclides, the identification of different aquifers as well as the dominant geochemical processes currently affecting the groundwater. The chemical characteristics of a groundwater can also furnish important constraints on any hydrologic model. When the physical characteristics of the hydrogeology are consistent with the chemical characteristics, the reliability of any predicted scenario is considerably greater.

Geochemical investigations of the Stripa groundwaters were initiated by Fritz, Barker and Gale [1] under the Swedish-American Cooperative (SAC) Program between SKBF/KBS and LBL. A considerable number of measurements were made during the SAC program from 1977-1980; these measurements have been discussed by Fritz and co-workers [1, 2, 3, and this symposium] and will not be elaborated on at this time other than to point out the main results.

These studies showed that, with increasing depth, the Stripa groundwaters (a) become more saline, (b) increase in pH up to about 10, and (c) decrease in alkalinity. It was also apparent that the Mg and K concentrations were anomalously low compared to most groundwaters. The salinity is primarily due to Na, Ca, Cl and SO_4. These results are somewhat unusual when compared to most groundwaters in granitic or metamorphic rocks, although analyses of water samples from these depths are almost non-existant in the published literature. The exponential increase in Cl with depth can be seen in Figure 1; this may be related to the decrease in hydraulic conductivity with depth. Three important questions remain largely unresolved from the previous studies at Stripa: What is the origin of the salinity? What are the trace element concentrations from the deep groundwaters? Can the trace elements help to identify the source of the salinity?

In 1981 a trace element program was initiated by the U.S. Geological Survey in cooperation with KBS and the Swedish Geological Survey. The purpose of this report is to present some of the results from this investigation and to provide further insight into the salinity problem of the Stripa groundwaters.

ANALYTICAL RESULTS

Water samples from shallow, intermediate and deep groundwaters at Stripa have been analyzed for 43 major and trace constituents. Two représentative analyses from the lower portion of V1 and V2 boreholes are shown in Table I. The elements Al, Mn, Cu, Zn, Co, Ti, As, Se, Sb, Pb, Tl, V, Bi, Cd, Ni, and Be are at or below detection. Analyses for NO_2, PO_4, and NH_4 showed them to be below detection and that both sulfide and Hg are extremely low. These results indicate that contamination from drilling equipment and other sources were probably minimal. Special sampling procedures and an interlaboratory comparison gave added assurance to the quality control (see details in Stripa Progress Report). The high quality of these results for major constituents is also reflected in the low cation-anion charge balance.

CHARACTERIZATION OF THE SALINE SOURCE

Chloride is often a good tracer in natural waters because it tends to be very non-reactive or "conservative" with respect to geochemical reactions. For this reason it is used to normalize chemical data and to distinguish conservative from non-conservative constituents. Analyses for Na and Ca in

TABLE I. - Typical Water Analyses from V1 and V2 Boreholes at Depth

Drillhole		V-2	V-1
Sample Code No.		81WA201	81WA203
Sampling Depth		763.7 - 877.7 m.	764.7 - 856.7 m.
Date Collected		3-6-81	3-6-81
Temperature (°C)		8.0	10.6
Field (Lab) pH		9.53 (8.69)	9.31 (7.57)
Conductivity (μS/cm)		960 (1415)	1420 (2230)
Eh (from field emf, mV)		+20 to +79	-56
Total Alkalinity (1)		11.5	9.25
Charge Balance (%)		0.18	1.1
Species			
Ca	mg/L	104	172
Mg	mg/L	0.11	0.19
Na	mg/L	180	277
K	mg/L	0.37	1.2
SO_4	mg/L	44.5	102
H_2S (2)	mg/L	0.00023	0.0026
F	mg/L	3.9	4.6
Cl	mg/L	410	630
Br	mg/L	4.0	6.5
I	mg/L	0.11	0.16
PO_4	mg/L	<0.1	<0.1
SiO_2	mg/L	12	13
B	mg/L	0.20	0.25
NO_2	mg/L	<0.005	<0.005
NO_3	mg/L	4.8	7.0
NH_4	mg/L	<0.01	<0.01
Al	mg/L	<0.005	0.008
Fe	mg/L	0.088	0.004
Mn	mg/L	0.009	<0.005
Cu	mg/L	<0.003	<0.003
Zn	mg/L	<0.005	<0.005
Hg	μg/L	0.023	<0.005
Co	mg/L	<0.005	<0.005
Ti	mg/L	0.014	0.035
Mo	mg/L	0.051	0.027
Li	mg/L	0.038	0.085
Sr	mg/L	0.99	1.7
Cs	mg/L	0.046	0.074
Rb	mg/L	0.015	0.032
Ba	mg/L	0.024	0.035
DOC (3)	mg/L	4.0	4.2

(1) Expressed as mg/L^{-1} HCO_3 equivalent.
(2) Values are minimums.
(3) Dissolved organic carbon.
(4) The following elements were found to be below detection limit for all samples (d.l. in ppm): As (0.003), Be (0.002), Bi (0.080), Cd (0.005), Ni (0.004), Pb (0.010), Sb (0.20), Se (0.20), Tl (0.007), V (0.005).

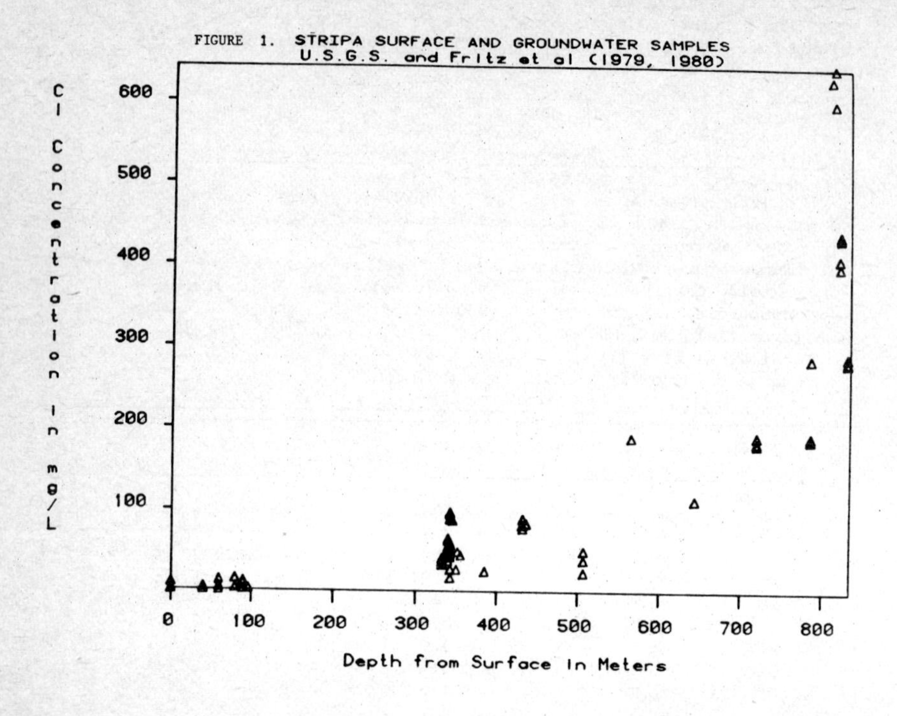

FIGURE 1. STRIPA SURFACE AND GROUNDWATER SAMPLES
U.S.G.S. and Fritz et al (1979, 1980)

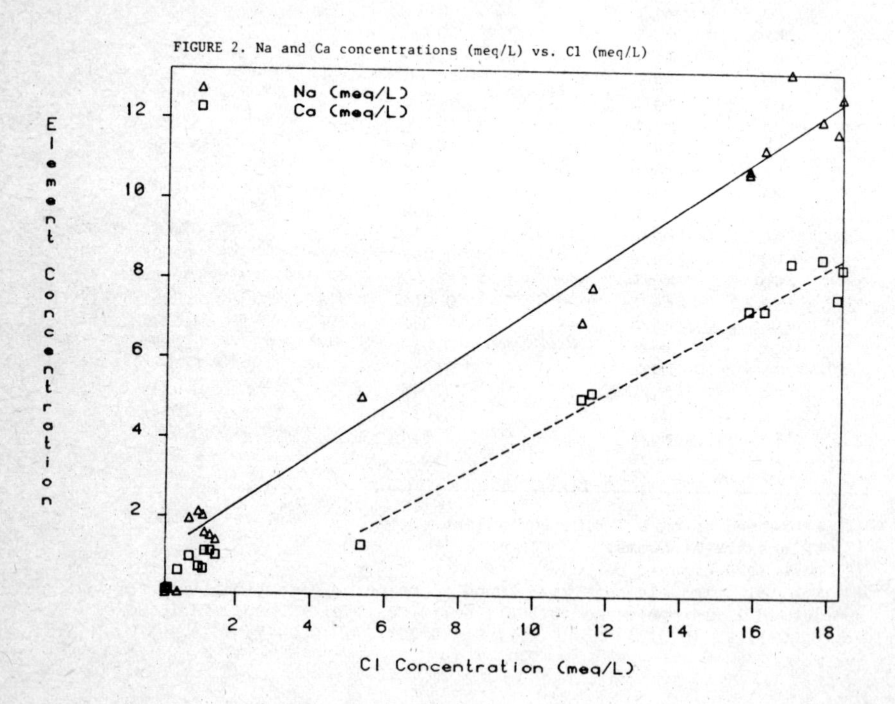

FIGURE 2. Na and Ca concentrations (meq/L) vs. Cl (meq/L)

groundwater samples collected during 1981 have been plotted against Cl (in meq/L) in Figure 2. At Cl concentrations greater than 5 meq/L, Na and Ca increase linearly with nearly parallel slopes. This plot shows the close correlation between the major contributors to the salinity. Bromide and iodide also correlate very strongly with Cl. These correlations would be expected from the mixing of two different waters: a dilute meteoric water with a Na-Ca-Cl brine (or the salts of such a brine). When SO_4 and HCO_3 are plotted vs. Cl (Figure 3) very non-conservative trends are seen. The HCO_3 concentrations decrease exponentially with increasing Cl which can be explained by calcite precipitation [1, 4]. The SO_4 concentrations, however, increase more rapidly than Cl. This rapid increase suggests an input of SO_4 to the groundwater by some source *in addition to* the saline source. Two obvious possibilities are pyrite oxidation and anhydrite (or gypsum) dissolution. Pyrite oxidation is highly unlikely because this reaction produces acid which would make the pH considerably lower. Anhydrite dissolution must be considered a real possibility.

Chemical data for the Stripa groundwaters can be generally grouped into conservative (Na, Li, Rb, Ca, Sr, Ba, Br, I, and NO_3) and non-conservative (K, Cs, Mg, HCO_3, F, SO_4, B, Mo, Al and Fe) constituents relative to Cl. This division does not mean that the conservative constituents do not participate in geochemical reactions (e.g. Ca certainly does participate in calcite precipitation) but rather that they are dominantly associated with a saline source; whereas the non-conservative constituents are dominantly controlled by water-mineral reactions. It should also be noted that the "conservative" constituents may participate in reactions to varying degrees, e.g. although Ba tends to increase with increasing Cl, it does so at a very low rate because of removal by coprecipitation with Ca in calcite and the precipitation of barite.

Chemical parameters which are known to be sensitive indicators of different types of saline waters can help to identify the origin of the deep groundwaters at Stripa. The main types of saline waters are:

1. Seawater (low temperature origin)
   - (a) Recent: seawater intrusion
   - (b) Ancient: formation waters, oil-field brines, etc.
2. Hydrothermal fluids (high temperature origin)
   - (a) Geothermal: hot springs, geysers
   - (b) Metamorphic: associated with active metamorphism
   - (c) Magmatic: associated with active magmatism
   - (d) Fluid inclusions: leakage into groundwater

Seawater is easily identified by the Br/Cl ratio which is a very conservative parameter unless considerable evaporation has taken place. Seawater intrusion into an aquifer will maintain a Br/Cl ratio of about 0.0034 and the relative amounts of some of the major ions such as Ca/Mg (=0.30) will not change drastically. The Br/Cl ratio for all samples collected from V1, V2 and N1 average 0.0104 ($\pm$ 0.0005) and the Ca/Mg ratio is more than two orders of magnitude higher than that of seawater. Modern seawater does not have the characteristics of the deep groundwaters at Stripa.

Seawater that has evolved over thousands of years can change significantly in chemical composition. The only type of saline water that occasionally has a Br/Cl ratio similar to the Stripa waters is oil-field waters. However, the geology at Stripa is quite far removed from that of an oil-field water in a sedimentary basin environment. More importantly, the Mg/Cl ratios are much lower and the Ca/Mg ratios are much higher in Stripa waters than in oil-field waters. Thus, even evolved ancient seawater has little in common with the saline waters from Stripa. Saline fluids of high temperature origin must be considered now.

Hydrothermal fluids are an unlikely source of salinity at Stripa because there is no indication of any thermal activity anywhere near the site.

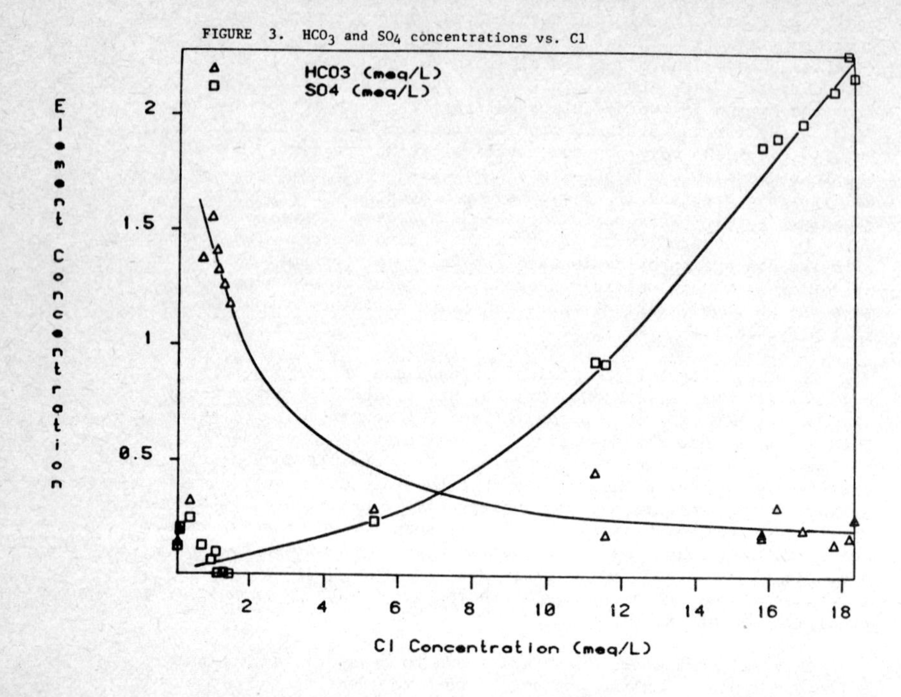

FIGURE 3. HCO_3 and SO_4 concentrations vs. Cl

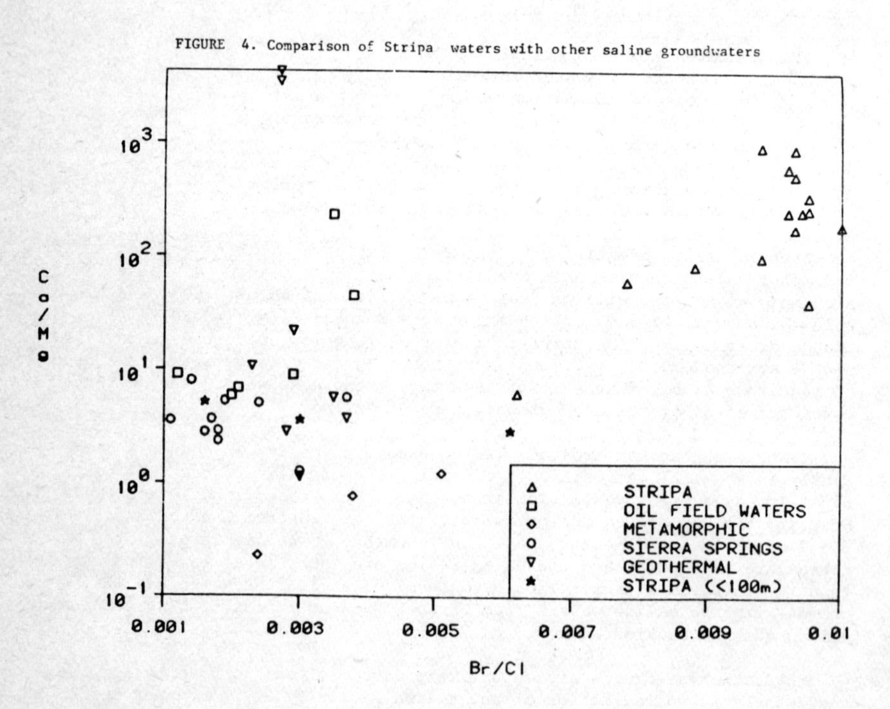

FIGURE 4. Comparison of Stripa waters with other saline groundwaters

In fact, the geothermal gradient is rather low (15°C/km). To compare the Stripa groundwaters with other saline waters which are typical of both high and low temperature origins, a plot of Ca/Mg ratio vs. Br/Cl ratio is shown in Figure 4. "Oil-field waters" are from the deep sedimentary basin of the Texas Gulf Coast [5], "geothermal" waters are from the Wairakei and Broadlands area, New Zealand [6], "metamorphic" waters are from the west coast of the United States [7], and "Sierra springs" are from the Sierra Nevada granitic batholith which are issuing from areas of active metamorphism [8]. For all of these types of saline waters, the deep Stripa groundwaters are clearly in a separate class from known saline waters of both high temperature and low temperature origins.

Shallow groundwaters at Stripa do have Ca/Mg and Br/Cl ratios similar to those of other saline waters, but the deep waters do not. It is noteworthy that the Ca/Mg ratio for oil-field and geothermal waters ranges over several orders of magnitude. This ratio is roughly correlated with temperature [9] such that the ratio increases at higher temperatures. The reason for this correlation is partly understood; it is caused by the marked decrease in Mg concentrations in high temperature fluids during the formation and equilibration of chlorite. An average Ca/Mg ratio from V1 and V2 drillholes gives a temperature of 520°C when plotted on the regression curve of Fournier and Potter [9]. Ellis [10] has shown that the $Mg/(H+)^2$ ratio is a good indicator of high temperatures, and applying his data to the deep groundwaters at Stripa results in temperatures around 250°C. Furthermore, the Mg/Cl ratio for the deep Stripa groundwaters, which is two orders of magnitude lower than the seawater value, compares very closely with geothermal waters. These temperatures of course, are based on the unconfirmed assumption that the analyzed Mg concentration is reflecting the Mg associated with the saline source. However, if some of the Mg is not associated with the saline source, the amount that is associated is less than the total; therefore the estimated temperatures are minimum values. Since no thermal anomalies are present at Stripa today, the possibility of leaching fluid inclusions from the crystalline rock must be considered as a possible source of the salinity. These fluid inclusions would be frozen-in memories of the last metamorphic event and would show some evidence of those high temperatures if (a) a high enough proportion of fluid inclusion brine has mixed with infiltrating meteoric water and (b) low temperature equilibration has not occurred.

The classical silica, Na/K and Na/K/Ca geothermometers [6, 11] are misleading for these groundwaters because of re-equilibration problems (assuming a high temperature saline fluid was present). The silica geothermometer, based on the large increase in silica solubility with temperature, is inappropriate if prolonged cooling has taken place. Silica will precipitate upon cooling; since the last thermal even at Stripa was probably several millions of years ago, the silica geothermometer would be useless. The Na/K geothermometer gives a temperature of -10°C which is indicative of high Ca concentrations and invalidates its use. The Na/K/Ca geothermometer gives a temperature of 61°C; this can only result from the association of the saline source with higher temperatures at some point in the geologic past or at some location near recharge because measured drillhole temperatures are only 8-12°C. The discrepancy between the Na/K/Ca temperature and the estimates based on Ca and Mg concentrations could be the result of low temperature equilibration of the alkalis in contact with feldspars.

To test the possibility that fluid inclusions in the granite might be the source of the groundwater salinity, nine samples of V1 and V2 drillcore were leached according to the procedures of Roedder, Ingram and Hall [12] and Hall and Friedman [13]. A total of 15 leaches from these samples gives an average Br/Cl ratio of 0.0101 ± 0.0015 compared to the average of 0.0104 ± 0.0005 from all analyses of V1, V2 and N1 water samples. This close comparison provides the strongest evidence to date that fluid inclusions can leach out of crystalline rocks at great depth. Groundwaters from E1 borehole give a lower Br/Cl ratio; this indicates a large input of shallow groundwater because surface water and shallow wells have much lower ratios.

The data from E1 gives additional evidence to the ^{4}He and ^{3}H data, indicating higher permeability in the E1 borehole with some connection to the surface.

The high Br/Cl ratios can also be related to a high temperature geologic event. If a fluid having the composition of seawater is heated during metamorphism, some of the chloride can substitute into phyllosilicates and other minerals containing hydroxyl sites. The higher the temperature, the more easily the Cl-OH substitution can take place. Bromide ion, however, cannot substitute because its radius is prohibitively large. Therefore, Br would be enriched in the fluid and Br/Cl ratios would increase.

FLUID INCLUSIONS AND SALINE GROUNDWATERS

Direct evidence has been presented for a possible connection between fluid inclusions and saline groundwaters occurring deep within the Stripa granite. This hypothesis is not new. For example, Garrels [14] mentions this as a possible source of chloride to groundwaters in igneous rocks; Gambell and Fisher [15] suggested that during rock weathering significant quantities of chloride may be released from crystalline rocks to account for the mass balance of chloride in the hydrologic cycle; Jacks [4] considered slow diffusion from fluid inclusions to be a possible source of chloride in the Stripa groundwaters and Fritz, Barker, and Gale [1] also list it as a possible source. In fact, the chemical data and element ratios obtained to date for the saline constituents are more consistent with an interpretation based on the washing out of fluid inclusions by recent meteoric water than with any other single hypothesis. Therefore, it is worthwhile to briefly consider the possible mechanism for the correlation between fluid inclusions and saline waters in crystalline rock.

Fluid inclusions commonly occur in planar structures within crystalline rocks, cutting across grain boundaries during the healing of fractures and microjoints [16]. The close relationship between fractures and fluid inclusion planes has been known for some time. Dale [17] notes instances where fluid inclusion planes and microfractures are parallel and coincident in granites and gneisses from New England. Examples of the same inclusion plane becoming a fracture has also been observed [17]. Since the hydrologic flowpaths in crystalline rocks are along permeable fractures, infiltrating groundwater has a very high probability of mixing with significant amounts of fluid inclusions. However, increases in groundwater salinity would only be apparent if the water to rock ratio was quite low and the ratio of infiltrating water volume to fluid inclusion volume was within a feasible range. These conditions would be optimal at a site such as Stripa where the hydraulic conductivity becomes very low at depths greater than 500 m. As long as the groundwater is limited to fractured zones and the permeability is so low, continued increases in salinity would be expected with depth.

If the salinity is derived from fluid inclusions, a few additional statements can be made. First, the limit of the increase in salinity would be the dissolved solids concentration of the fluid inclusion brine. We have been able to estimate the water content along with the chloride content of several drillcore samples of V1 and V2 which give concentrations of 50-150 g/L Cl. These concentrations must be considered highly uncertain because the water content is extremely difficult, if not impossible, to measure for small inclusions (E. Roedder, personal communication). Nevertheless, reliable measurements of fluid inclusion brines would suggest that these concentrations are very common. Crystals of halite are commonly found in fluid inclusions; so it is quite feasible for groundwater in fractured systems to reach compositions approaching halite saturation. Frape and Fritz [18] have found saline groundwaters in the Canadian Shield which are as high as 156 g/L and these may represent a residual metamorphic fluid. It is also noteworthy that seven analyses for ^{2}H from water removed from V1 and V2 drillcores gave values in the range of -52 to -86 $^{o}/oo$ (relative to SMOW) which are exactly in the same range as the Canadian Shield groundwaters. Thus, if the extracted water is from fluid inclusions, these data would indicate that the ^2H in the groundwaters would start increasing once the proportion of water from fluid inclusions reached about 5-10% of the total water in the fracture zone.

Another aspect of the chemistry that bears on the hydrogeology is the sensitivity of the Br/Cl to change with location and depth in the subsurface. All of the deep groundwaters from V1 and V2 consistently have ratios of about 0.010 which is consistent with the same ratio found in the Stripa granite. Water from borehole N1, although horizontal, maintains the same Br/Cl ratio indicative of older, more slowly moving groundwaters. In contrast, E1 has lower Br/Cl ratios (0.006) which are more indicative of surface and shallow groundwaters. The implication is that E1 must have zones of greater permeability to the surface than N1. Furthermore, the crushed and highly permeable zone in V1 does not show signs of shallow water influx. If the Br/Cl ratio in the water from the crushed zone in V1 decreases with time, a connection to shallow groundwaters would then be very likely.

NON-SALINE CONSTITUENTS

Several constituents, especially F, SiO_2, Mg and Mo, bear no relationship to increasing Cl concentrations. The fluoride concentrations are likely to be limited by the solubility of fluorite which has been shown to be an effective control in geothermal waters [6] and low temperature groundwaters [4, 19]. Mineral saturation calculations obtained with WATEQ2 [20] consistently give slightly supersaturated conditions with respect to fluorite and calcite, suggesting that both of these minerals are actively precipitating. The driving force for precipitation is the increasing Ca concentrations associated with the saline source.

Silica concentrations are remarkable low for these pH values of 9-10. As the pH increases above 9.5, the dissolved silica would increase to 20 or 30 mg/L if if were in equilibrium with microcrystalline silica such as chalcedony. Since it doesn't, a solubility control by complex aluminosilicates must be considered. Concentrations of Mg are so low that it is effectively a trace element. These trace concentrations are only possible in the absence of any significant amounts of bicarbonate and any Mg-bearing saline waters. These conditions, which occur at Stripa, may indicate the equilibrium saturation of chlorite which is abundant in the fracture minerals. Too little is known about the occurrence of Mo in groundwaters to make any comment except that it occurs chiefly as an oxyanion in the dissolved state, like silica, and that it might be expected to behave somewhat similarly.

CONCLUSIONS

An examination of the Stripa groundwater chemistry has shown several anomalous trends that do not correlate well with most other types of groundwaters. Chemical profiles with depth, which show increasing salinity, increasing pH and decreasing alkalinity, indicate mixing of meteoric water with a Na-Ca-Cl type water. The Br/Cl ratio of the deep groundwaters is very high compared to most other saline groundwaters. However, this ratio does match that found by leaching fluid inclusions from the Stripa granite. This evidence suggests that the salinity is caused by leaching of fluid inclusions from fractures in the granite and associated metamorphics. Element ratios, especially Ca/Mg, indicate that the salinity has a high temperature origin which is consistent with the leaching of fluid inclusions from igneous and metamorphic rocks. Further research on deep groundwaters in igneous and metamorphic rocks needs to investigate the fluid inclusion composition of these rocks to confirm this hypothesis and to determine if this phenomenon is widespread or relatively rare.

REFERENCES

[1] Fritz, P., Barker, J.F. and Gale, J.E. : "Geochemistry and Isotope Hydrology of Groundwaters in the Stripa Granite: Results and Preliminary Interpretation", LBL-8285, SAC-12, 105 pp. (1979).

[2] Witherspoon, P.A., Cook, N.G.W. and Gale, J.E. : "Geologic Storage of Radioactive Waste: Field Studies in Sweden", Science 211, 894-900 (1981).

[3] Andrews, J..N., Giles, I.S., Kay, R.L.F., Lee, D.J., Osmond, J.K., Cowart, J.B., Fritz, P., Barker, J.F., and Gale, J.E. : "Radioelements, Radiogenic Helium, and Age Relationships for Groundwaters from the Granites at Stripa, Sweden", Geochim. Cosmochim. Acta, in press (1982).

[4] Jacks, G. : "Ground Water Chemistry at Depth in Granites and Gneisses", KBS Tech. Rept. 88, 28 pp. (1978).

[5] Kharaka, Y.K., Callender, E. and Carothers, W.W. : "Geochemistry of Geopressured Geothermal Waters from the Texas Gulf Coast", Proc. Third Geopress Geotherm. Energy Conf., John Meriwether, ed., Vol. 1, 121-165 (1977).

[6] Ellis, A.J. and Mahon, W.A.J. : Chemistry and Geothermal Systems, Academic Press, N.Y., 392 pp. (1977).

[7] Barnes I. : "Metamorphic Waters from the Pacific Tectonic Belt of the West Coast of the United States", Science 168, 973-975 (1970).

[8] Barnes, I., Kistler, R.W., Mariner, R.H. and Presser, T.S. : "Geochemical Evidence on the Nature of the Basement Rocks of the Sierra Nevada, California", U.S. Geol. Survey Water-Supply Paper 2181, 13 pp. (1981).

[9] Fournier, R.O. and Potter, R.W. : "Magnesium Correction to the Na-K-Ca Chemical Geothermometer", Geochim. Cosmochim. Acta 43, 1543-1550 (1979).

[10] Ellis, A.J. : "Magnesium Ion Concentrations in the Presence of Magnesium Chlorite, Calcite, Carbon Dioxide, Quartz", Am. J. Sci. 271, 481-489 (1971).

[11] Truesdell, A.H. : "Geochemical Techniques in Exploration", in Proc. Second U. N. Symp. Development Use Geotherm. Resour., Vol. 1, liii-lxxix (1975).

[12] Roedder, E., Ingram, B. and Hall, W.E. : "Studies of Fluid Inclusions III: Extraction and Quantitative Analysis of Inclusions in the Milligram Range", Econ. Geol. 58, 353-374 (1963).

[13] Hall, W.E. and Friedman, I. : "Composition of Fluid Inclusions, Cave-in-Rock Fluorite District, Illinois, and Upper Mississippi Valley Zinc-Lead District", Econ. Geol. 58, 886-911 (1963).

[14] Garrels, R.M. : "Genesis of Some Ground Waters from Igneous Rocks", in Researches in Geochemistry, P. H. Abelson, ed., Vol. 2, 405-420 (1967).

[15] Gambell, A.W., and Fisher, D.W., "Chemical Composition of Rainfall in Eastern North Carolina and Southeastern Virginia", U.S. Geol. Survey Water-Supply Paper 1535-K, 41 pp. (1966).

[16] Wise, D.U. : "Microjointing in Basement, Middle Rocky Mountains of Montana and Wyoming", Geol. Soc. Am. Bull. 75, 287-306 (1964).

[17] Dale, T.N. : "The Commercial Granites of New England", U.S. Geol. Survey Bull. 738, 14-26 (1923).

[18] Frape, S.K. and Fritz, P. : "The Chemistry and Isotopic Composition of Saline Groundwaters from the Sudbury Basin, Ontario", Can. J. Earth Sciences 19, 645-661 (1982).

[19] Handa, B.K. : "Geochemistry and Genesis of Fluoride-Containing Groundwaters in India", Groundwater, 13, 275-281 (1975).

[20] Ball, J.W., Jenne, E.A., and Nordstrom, D.K. : "WATEQ2 - A Computerized Chemical Model for Trace and Major Element Speciation and Mineral Equilibria of Natural Waters", in Chemical Modelling in Aqueous Systems, E.A. Jenne, ed., Am. Chem. Soc. Symp. Series 93, 815-836 (1979).

ACKNOWLEDGEMENTS

I am very grateful for the technical and interpretive assistance of Jim Ball and Rona Donahoe. This investigation was carried out by a cooperative agreement between SKBF/KBS and the U.S. Geological Survey under the auspices of OECD/NEA. Discussions with Peter Fritz, Blair Jones, Ed Roedder and Ivan Barnes have been most helpful, although they are absolved of any responsibility for the views presented.

MIGRATION EXPERIMENTS IN A SINGLE FRACTURE IN THE STRIPA GRANITE.

PRELIMINARY RESULTS.

Harald Abelin, Jard Gidlund, Ivars Neretnieks
Department of Chemical Engineering
Royal Institute of Technology
S-100 44 STOCKHOLM, Sweden

SUMMARY

Within the Stripa project flow and sorption in a readily identifiable fracture which can be excavated for a detailed examination of the flow path and sorption sites is under investigation. The experiments are performed in the Stripa mine, 360 m below ground, where there is a natural water flow towards the drift. The bedrock is granite.

A method of tracer injection into a fracture, either as a step or a pulse, and of collection of water samples under anoxic atmosphere has been tested in a preparatory investigation. The introduction of tracers is done by injection with over pressure.

Injections of Rhodamine-WT, Na-iodide and Uranine have been performed. It has been found that Rhodamine-WT is influenced in some way along the flow path.

Another fissure system has been chosen for the main test. In this sorbing tracers will also be used. So far tests for connectivity and a tracer test using Uranine has been made.

1. BACKGROUND

In the KBS (Swedish nuclear fuel supply co/division KBS) report [1], it is proposed that the final repository for radioactive waste should be at 500 m depth in crystalline rock. The safety analysis for this repository is based on the assumption that if and when any radionuclides are leached from the waste, the majority of the important radionuclides will interact chemically or physically with the bedrock and will be considerably retarded. This retardation and interaction depends upon the velocity of water, the sorption rates and equilibria of the reactions as well as the surface area of the rock in contact with the flowing water.

Most studies are based upon the assumption that the flow can be described as flow in a porous medium. This might be true for very large distances where the flow would encounter a multitude of channels and some averaging may be conceivable on the scale considered. However, no large scale tracer tests have been performed in fissured crystalline rock with known flow paths. Transport over short distances, i.e. in the near field of a canister, most probably occurs in individual fissures. On an intermediate scale where more than a few fissures conduct the flow, well type tracer tests alone cannot give the detailed information needed to understand dispersion and sorption phenomena in fissured rock. It has therefore been decided to investigate flow and sorption in readily identifiable fissures which can be excavated for a detailed examination of flow paths and sorption sites.

Several investigations on migration in single fractures are under way both in the laboratory and in the field. The laboratory runs are done with migration distances of up to 0.3 m (Neretnieks et al. [2]) and the field experiments with migration distances of up to 30 m (Gustavsson and Klockars [3]). In the present investigation a migration distance of about 5 m is used. The fracture will be excavated afterwards.

2. PURPOSE

The study has the following main objectives:

- To observe the movement of nonsorbing and sorbing tracers under controlled and well defined conditions in a real environment.
- To interpret the movement of tracers in such a way that the results become useful for the prediction of radionuclide migration.
- To obtain a basis for comparing laboratory data on sorption with observations in a real environment.
- To develop techniques for small volume sampling of water and fracture surfaces with sorbed tracers.
- To gather experiences with stable tracers before using radioactive tracers.

3. EXPERIMENTAL DESIGN

This investigation started in 1980 and will end in late 1983. Injection of sorbing tracers will start in October 1982.

The Stripa mine is well suited for performing tracer tests in a single fracture as well as in a network of interconnected fractures. Old water bearing fractures have been found in and near the drifts now in use. As the drifts are well below the water table, the fractures have been conducting water for a very long time and thus are as well "equilibrated" as we can reasonably achieve in a sorption experiment.

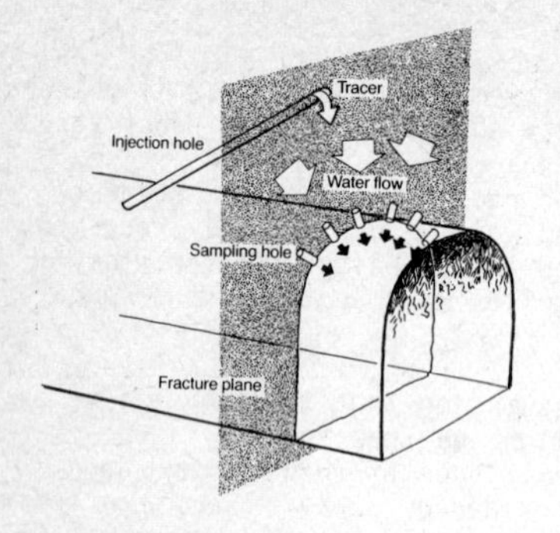

The technique used in this investigation is to locate a suitable fracture, drill an injection hole which intersects the fracture at a distance of about 5 m from the face of the drift, and several sampling holes in the fracture plane (see fig 1). Tracers are introduced into the fracture from the injection hole by injection with an over pressure. The natural water flow is towards the drift. Groundwater with tracers is collected in sampling holes. By having a series of sampling holes the transverse dispersion of the tracers can be observed in addition to the axial dispersion.

Figure 1. Drift with intersecting-fracture.

3.1 Injection and sampling techniques

The injection hole is sealed off just below and above the fracture with two straddled mechanical packers, see figure 2 (In the preparatory investigation an inflatable packer was used. In order to reduce the injection compartment volume a PVC filling with a diameter near that of the borehole was used.) In the main investigation the injection compartment volume is minimized by making the injection section, which lies between the two straddled packers, nearly the same diameter as the hole. The introduction of tracers in the fracture was made by injecting the tracers with a certain over pressure. The injected volume and the injection flow is measured.

Each sampling hole has a mechanical packer (see figure 2), with a funnel shaped top. It is possible to purge the sampling holes with nitrogen, in order to maintain the redox integrity of the samples. The water from the sampling holes is collected with a fractional collector which can be kept under an anoxic atmosphere.

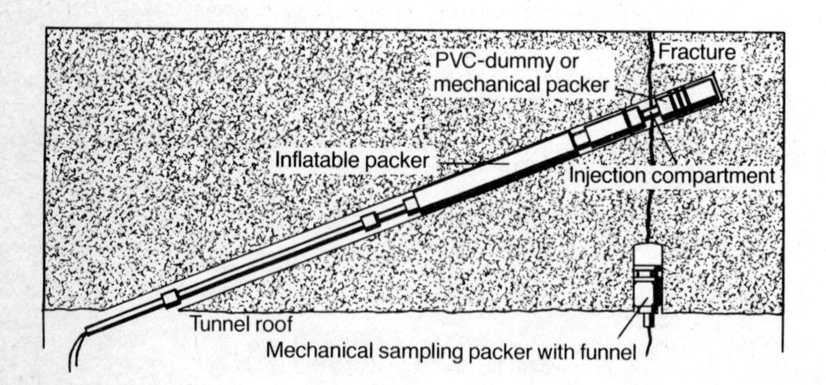

Figure 2. Injection and sampling packers

A more detailed description of the equipment is given in Abelin and Neretnieks [4].

3.2 Tracers

Stable tracers are used throughout in the field experiments. Except for very few nuclides, notably Np, Pu and Tc all the objectives stated can be achieved with less effort. "Hot" laboratory experiments are run in parallel in a supporting investigation, Eriksen [5].

Four classes of tracers were considered:

o Particles	0.4 micrometer plastic pellets were tried to simulate the movement of particulate matter. No suitable method to analyse the particles was found and they will not be used.
o High molecular weight tracers	Blue Dextran M = 2 000 000 or Albumine is used to simulate the movement of high molecular weight organic nonsorbing matter.
o Nonsorbing tracers	Various dyes, Bromide, Iodide.
o Sorbing tracers	Cs (I), Sr (II), Eu Nd (III), Th (IV), U (IV,VI).

The nonsorbing tracers are tested for sorption on crushed granite and materials from the equipment as well as stability in time. The sorption properties of 7 "nonsorbing" tracers have been determined. Eosin gelblich, Ebenyl Br. Flavine and Uranine showed negligible sorption on either granite or the materials used in the equipment. Amidoflavin FFP, and the three o, m, p - fluoro-benzoic acids either sorbed or displayed instabilities which make them less suitable as tracers in our system.

The analysis of the high molecular weight tracers, and the nonsorbing tracers are done by spectrophotometer, ion selective electrode, and atomic absorbtion analysis. The sorbing tracers in water and granite will be analysed by neutron activation analysis.

4. PREPARATORY INVESTIGATION

To test the methods for injection and sample collection a preparatory investigation was run. A suitable fracture for this was found at the 360 m level in the Stripa mine in the so called "ÖV1" drift.

Figure 3 shows the site for the preparatory investigation with its three main equipment components:
1. Injection equipment
2. Packer and sampling devices
3. Fractional collector with anoxic box

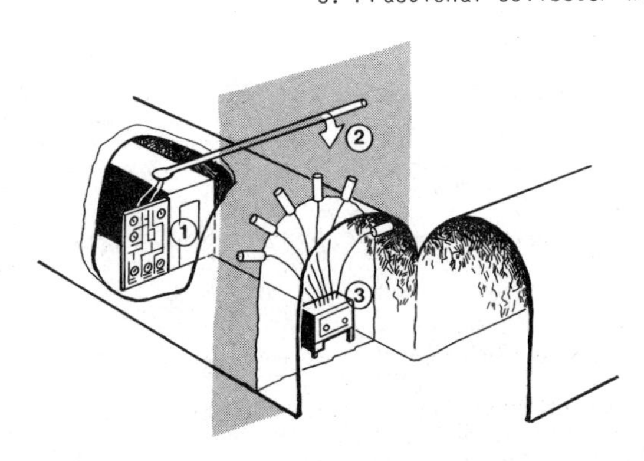

Figure 3: Test site at the 360 m level

A more detailed description of the equipment used in the preparatory investigation is given in [4].

4.1 Data on water flow, pressure and test for connection

After the injection hole was drilled, the water flow from the injection hole was monitored. A small mechanical packer was inserted near the end of the injection hole. The water flow was found to be approximately 100 ml/h. The natural pressure in the injection hole was found to be 0.28 MPa.

Table I. Water flow rate from sampling holes

Hole no	Water flow rate Q_r ml/h
1	0 (approx.)
2	0.2
3	5
4	83
5	28
6	0

The sampling holes are at a distance of about 0.7 m from each other and situated in the same fracture. Table I shows that there is a considerable channeling in the fracture. After the water flow was monitored, Uranine was injected into the fracture to test for connection between the injection hole and the sampling holes. It took less than 20 h for the first of the tracer to reach sampling holes no. 4 and no. 5. The injection was continued and tracer eventually arrived in hole no. 3 as well. No tracer was found in the remaining three holes. After this test, flushing of the fracture with groundwater was attemted, however, the tracer could be detected in the flowing water for a very long time.

4.2 Simultaneous run with Uranine and Rhodamine-WT

Uranine and Rhodamine-WT were injected simultaneously for 90 h. Thepressure used was 0.5 MPa. The results from this test run are shown in figures 4 and 5. Figure 4 shows the breakthrough curves for hole no. 4 and no. 5. It can be seen that Rhodamine-WT appears later and more diluted in both holes. Figure 5 shows the difference in breakthrough time and dilution between holes no. 4 and no. 5.

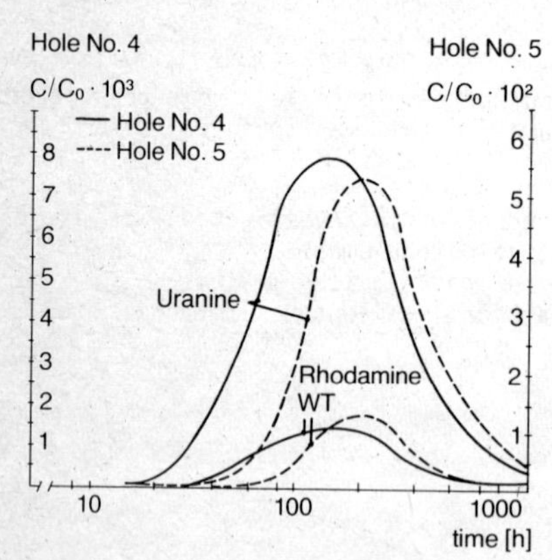

Figure 4. Breakthrough curves for holes no. 4 and no. 5.

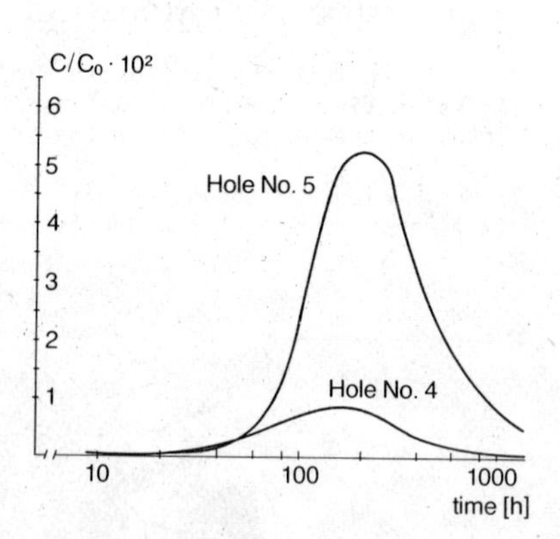

Figure 5. Comparison of breakthrough curves for holes no. 4

4.3 Tentative interpretation of the preparatory investigation

The very different flow rates to the various collecting holes indicate a considerable channeling. The mean travel time for Uranine is about 100 and 200 h for holes no. 4 and 5 respectively. From this, the known flow rates in the collection holes (83 and 28 ml/h), the collection length along the perimeter (about 0.7 m), and hydraulic head and geometry, the hydraulic conductivity K_{pf} and equivalent fracture width for flow d_f can be determined.

The hydraulic conductivity of the fissure

$$K_{pf} = 0.5 \ln(\frac{r_2}{r_1}) (r_2^2 - r_1^2)/t_w (h_2 - h_1) \quad (1)$$

The velocity of the wall of the drift (at r_1) is

$$v_{r1} = 0.5 \frac{1}{t_w} (\frac{r_2^2}{r_1} - r_1) \quad (2)$$

The equivalent fracture width for flow is

$$d_f = Q_r/l_{r1} v_{r1} \quad (3)$$

An equivalent fracture width for laminar flow in a parallel walled fracture with the same hydraulic conductivity can also be determined:

$$d_\ell = \sqrt{K_{pf} 12 \nu/g} \quad (4)$$

The following data were used in the calculation:

The radii are $r_1 = 2.25$ m and $r_2 = 6$ m.
The hydraulic heads are $h_1 = 0$ m and $h_2 = 28$ m.
The collecting length $l_{r1} = 0.7$ m.

The results are shown in table II below.

Hole no	K_{pf} m/s	d_f m	d_ℓ m
4	$1.4 \cdot 10^{-6}$	$1.8 \cdot 10^{-3}$	$0.0013 \cdot 10^{-3}$
5	$0.7 \cdot 10^{-6}$	$1.2 \cdot 10^{-3}$	$0.001 \cdot 10^{-3}$

5. MAIN INVESTIGATION

For the main investigation where in addition to nonsorbing tracers sorbing tracers will be injected, another site was chosen. This is located about 500 m from the site for the preparatory investigation. At the new site 6 injection holes have been drilled. Four of these intersect the main fracture at a distance of about 5 m from the face of the drift and the fifth intersects at about 10 m. One hole had to be plugged.

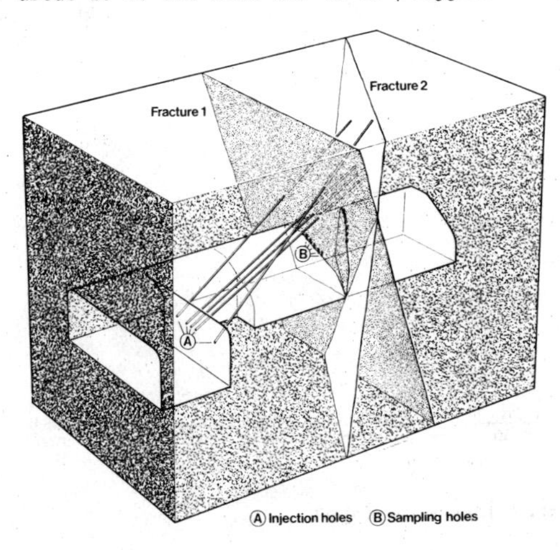

Figure 6. Main test site

Figure 6 shows the general layout of the site. The location of the connecting fissures is beeing determined by hydraulic pulse testing. Good connection has already been found with good accuracy from injection hole N2 to the surface of the drift. At the surface of the drift 30 sampling holes have been drilled. 20 holes are equipped with collecting devices, 11 of these carry water in measurable quantities. An example of the pressure responses in samling hole S2-8 with packer at various depths in the injection hole N2 is shown in figure 7. The loss of fluid in the injection hole during the constant pressure phase of the injection is shown in figure 8.

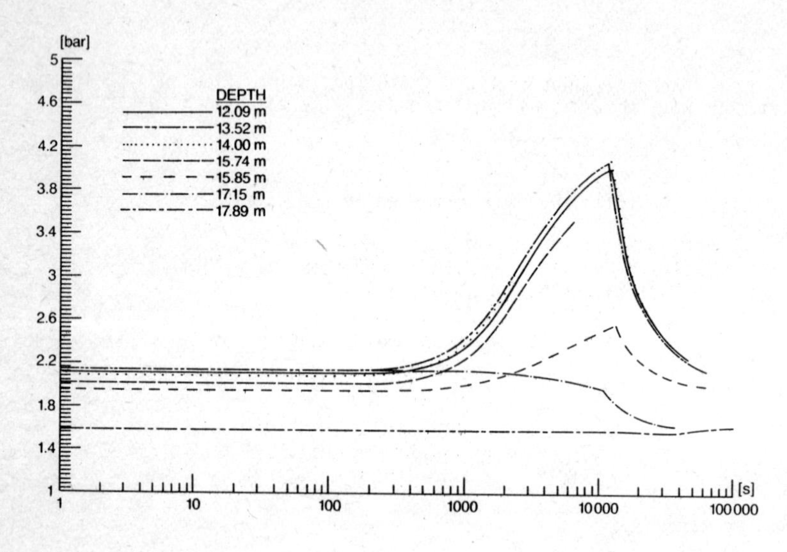

Figure 7. Pressure response in S2-8.

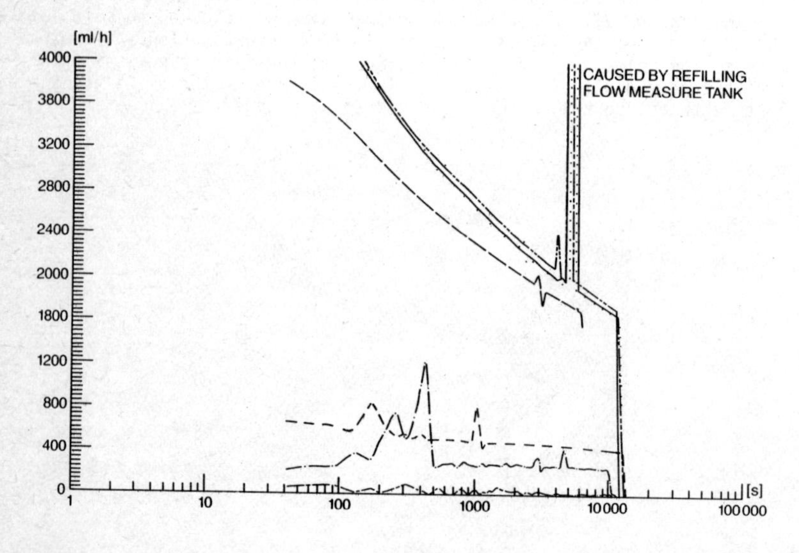

Figure 8. Water loss rate in N2.

The location of the fissure in the other holes is sought by the same pressure pulse technique by moving a single packer outward from the bottom of the hole by increments. The increments are determined from suspected fracture locations seen on the core and from TV logs.

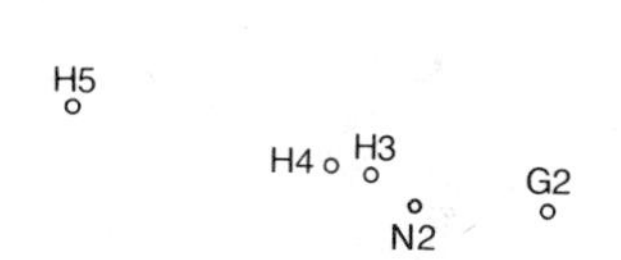

The location of the fissure intersects with the injection holes have been determined in 3 holes at present. One tracer test has been performed with injection in hole N2 and collection in 4 of the sampling holes. The two sampling holes S2-6 and S2-8 carried tracer, see figure 9. All other injection holes were sealed during this test and held slightly above normal hydrostatic pressure which is 0.2 MPa.

The flow rates in the sampling holes carrying most water are shown in table III. Approximate mean residence times are also given in the same table. The injection pressure was 0.03 MPa above the ambient pressure. The injection was maintained for 142 hours. The results are shown in figure 10.

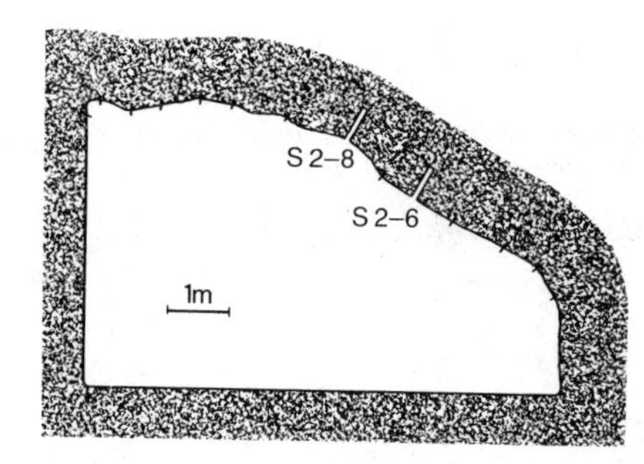

Figure 9. Location of injection and sampling holes in fracture 2

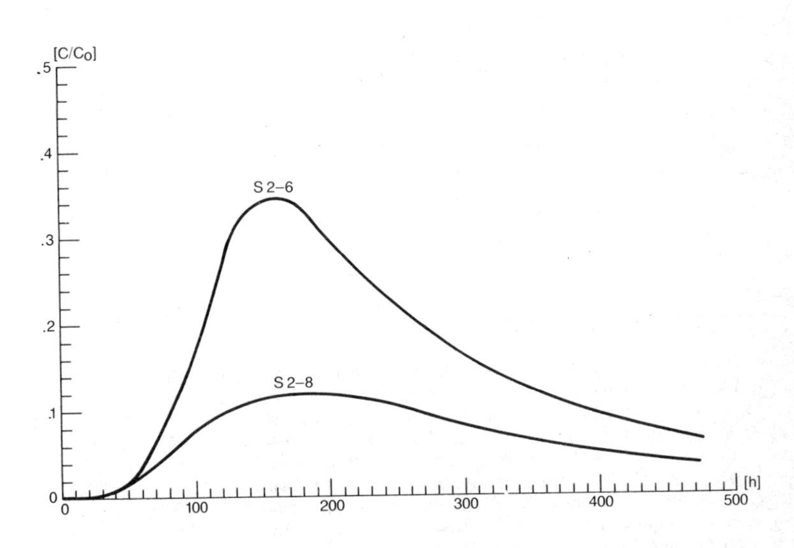

Figure 10. Breakthrough curves for sampling holes S2-8. S2-6

Table III. Water flow rate from sampling holes averaged over 26 days.

Hole no.	Water flow rate Q_r ml/h	Residence time t_w approximate hours
S 2.2	0.25	-
S 2.6	2.2	80
S 2.8	35.0	80
S 2.10	0.8	-

The same method to evaluate the fissure widths as was used for the preparatory test equations 1-4 is used.

The following data were used in the calculations
r_1 = 4 m r_2 = 8 m
h_1 = 0 m h_2 = 23 m
l_{r1} = 0.7 m

The radius r_1 is approximated as the mean of the width and height of the tunnel + the depth of the collection hole. l_{r1} is taken to be the distance to the next collection hole (which is not water bearing). Using those approxiamations the tentative results are shown in table IV.

Table IV. Fracture widths and hydraulic conductivities in main fissure 2.

Hole no.	K_{pf} m/s	d_f m	d_ℓ m
S 2-6	$2.6 \cdot 10^{-6}$	$0.042 \cdot 10^{-3}$	$0.0018 \cdot 10^{-3}$
S 2-8	$2.6 \cdot 10^{-6}$	$0.67 \cdot 10^{-3}$	$0.0018 \cdot 10^{-3}$

6. SUPPORTING LABORATORY INVESTGATION

Sorption for Cs and Sr and diffusion data have been determined for Stripa granite by Skagius et al. [6]. Cores from Stripa with longitudinal natural fissures have been used in "column tests" [2]. Nonsorbing (THO) as well as sorbing tracers (Cs, Sr) have been used. The breakthrough curves have been interpreted using a model which includes channeling, surface sorption, diffusion into the rock matrix and sorption on the inner surfaces of the rock. The agreement was good between predicted and experimentally obtained breakthrough curves.

7. DISCUSSION AND CONCLUSIONS

In the site for the main investigation the main connection is not always where it is expected from geometrical projection of the visible fissure in the drift and core and TV observations. Even in injection holes as near as 0.5 m from each other the good connection may not be found where expected. This indicates that the pathways are considerably more complex than "a single fissure". The pressure, flow and tracer observations indicate that there is a large difference in fissure width expected from pressure drop considerations in parallell wall flow and actual fissure widths. The actual fissure widths are considerably larger.

8. NOTATION

d_f	fracture width for flow	m
d_ℓ	equivalent fracture width from parallell plate theory	m
g	gravitational constant	m/s^2
h	pressure head	m
K_{pf}	hydraulic conductivity of the fracture	m/s
l_{r1}	collecting length at r_1	m
Q_r	water flow rate at r	m^3/s
r	radial distance from center of drift	m
t_w	water residence time	s
v_{r1}	water velocity at r_1	m/s
ν	viscosity of water	Ns/m^2

9. REFERENCES

1. "Handling of Final Storage of Unreprocessed Spent Nuclear Fuel", Vol II Technical KBS Report (1978).
2. Neretnieks, I., Eriksen, T., Tähtinen, P., "Tracer Movement in a Single Fracture in Granitic Rock. Some Experimental Results and their Interpretation", Water Resources Research in print 1982.
3. Gustavsson, G., Klockars, K-E., "Studies on Groundwater Transport in Fractured Crystalline Rock under Controlled Conditions using non Radioactive Tracers", Technical Report KBS 81-07.
4. Abelin, H., Neretnieks, I., "Migration in a Single Fracture. Preliminary Experiments in Stripa", Internal Report KBS 81-03.
5. Eriksen, T., Dep. Nucl. Chem., Royal Inst. Techn. Stockholm, Personal communic. 1982.
6. Skagius C., Svedberg G,, Neretnieks I., "A study of strontium and cesium sorption on granite. Nuclear Technology in print 1982.

GEOCHEMISTRY AND ISOTOPE HYDROLOGY OF GROUNWATER AT AURIAT

S. Derlich* and J.L. Michelet**
Commissariat à l'Energie Atomique
Institut de Protection et Sûreté Nucléaire - CSDR
Fontenay-aux-Roses - France

RESUME

Les eaux du forage profond d'Auriat (France) ont été étudiées par les méthodes de la géochimie et des isotopes du milieu.

Ces analyses montrent que les eaux sont un mélange d'eaux superficielles et d'eaux de foration.

La présence d'eaux de formation correspondant à une circulation réellement profonde n'a pu être mise en évidence à ce stade de l'étude.

ABSTRACT

Waters from deep borehole of Auriat (France) were investigated using geochemical and environmental isotopes techniques.

Analyses show that these are drilling waters mixed with shallow groundwaters.

Intraformational waters corresponding to really deep flow have not been evidenced at that stage of the investigation.

* Commissariat à l'Energie Atomique - IPSN-CSDR

** Université de Paris-Sud, Laboratoire d'Hydrologie et de Géochimie Isotopique

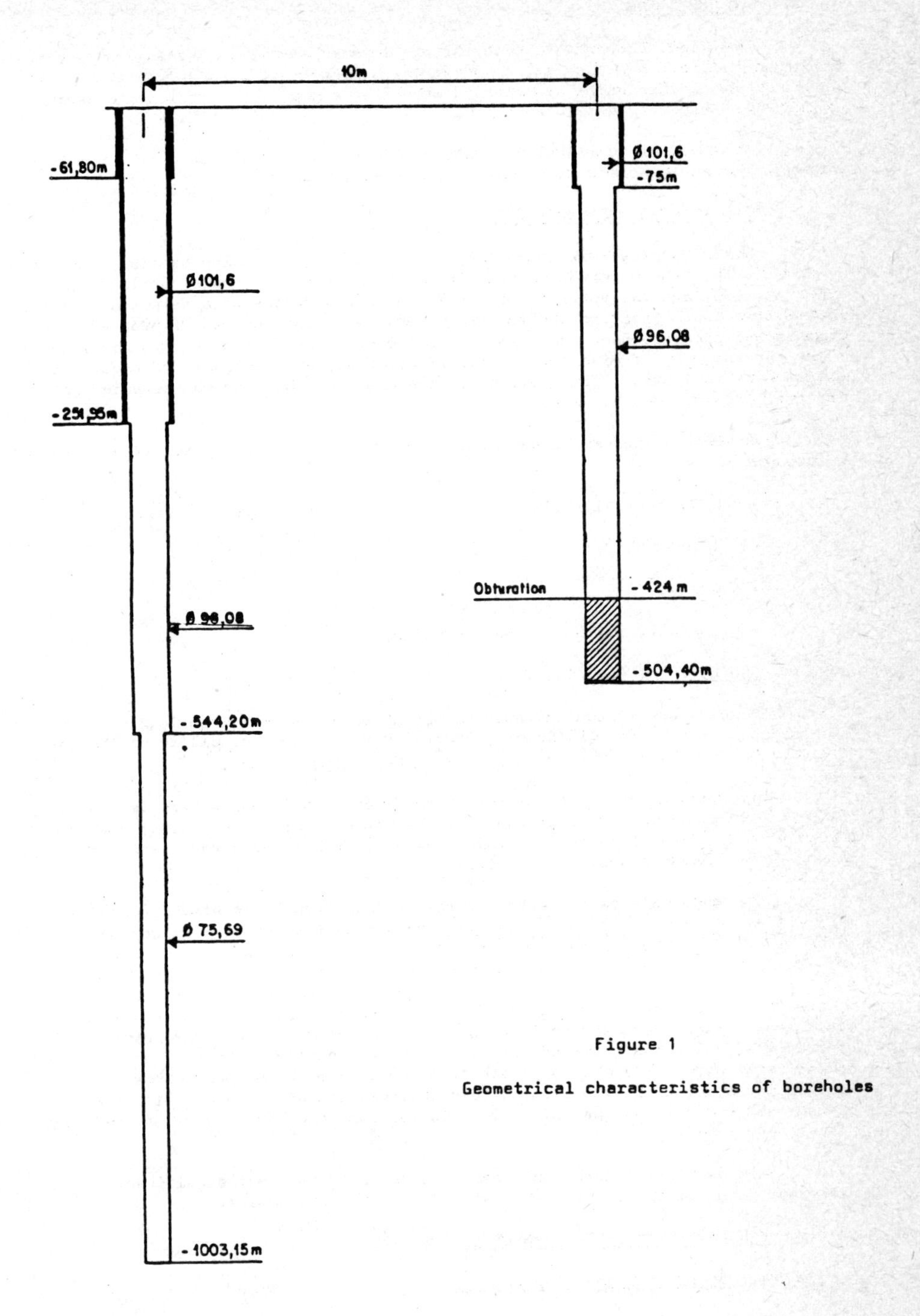

Figure 1

Geometrical characteristics of boreholes

1. INTRODUCTION

The first french test site for preliminary studies on geological waste isolation is located at Auriat, approximately 30 km NW of Limoges, on the western border of Massif Central. The selected massif is a granitic batholite, clearly delimited on about 92 km2 in area (1).

In 1980, two vertical boreholes (1000 m and 500 m) were cored (complete coring). Horizontal distance between the two boreholes is 10 m (fig.1).

2. SAMPLING AND MEASUREMENTS

Water samples were collected at different depths using special sampling devices in the 1000 m borehole. Sampling took place at the end of drilling operations (1980) and two years later (1982). Sampling depths were selected on basis of the core-logs (fracture and fissure zones), and from results of hydraulic and geophysical logging. Some samples were collected at depth pressure in order to allow the recovery of dissolved gases. In addition, some surface and superficial waters were collected : four springs in the area and the pond who supplied water for drilling operations.

Because of the small amount of available samples ($<$ 1 l.) it was decided to measure :

- pH, conductivity (field determination), major ions and silica.
- ^{18}O and ^{2}H.
- Tritium.
- Fe, Al, F, Sr, NO_3 in some samples.

3. GEOCHEMISTRY (samples 1980)

Results are listed in table 1. Spring waters show low pH (4.4. to 5.6) and low conductivity ($<$ 48 μScm^{-1}), according to the low solubility of the rock. They are HCO_3^- - Na^+ - Mg^{2+} or HCO_3^- - Na^+ - Ca^{2+} type waters.

For comparison, conductivity of pond water (drilling water) is lower (34.5 μ Scm^{-1}) and pH is higher than in subsurface waters (equilibrium with atmospheric CO_2 and biologic activity). However, in both grounwater and pond water, ionic contents are quite similar.

The chemical load of waters collected in borehole is strong (conductivity 407 and 422 μ Scm^{-1}). HCO_3^- and Na^+ are dominating ions. pH values are close to 8. Waters from the borehole show a rather high Cl^- content ($\simeq$ 20ppm). Further more equilibrium calculations (WATEQ F program) indicates supersaturation with respect to most of clay minerals (illite, kaolinite, Ca montmorillonite) and to K mica (2). These properties may suggest a remote circulation from sedimentary rocks and a rather long chemical evolution in contact with the rock matrix. Concentrated solutions (chloride rich) have been recently recorded from granitic batholite at Stripa (3) and in crystalline rocks from the Canadian Shield (4). However the saturation with respect to smectites strongly suggests also that a contamination due to the use of bentonitic mud during coring operations is possible.

Ratios of dissolved noble gases (e.g. Ar/N_2) are not significantly different from atmospheric values : waters are not paleowaters.

4. ENVIRONMENTAL ISOTOPE ANALYSES (table3)

4.1 Stable isotopes (^{2}H and ^{18}O)

Deuterium contents are plotted against oxygen 18 contents in figure 2. Springs waters lie close to the present day global meteoric water line (MWL)

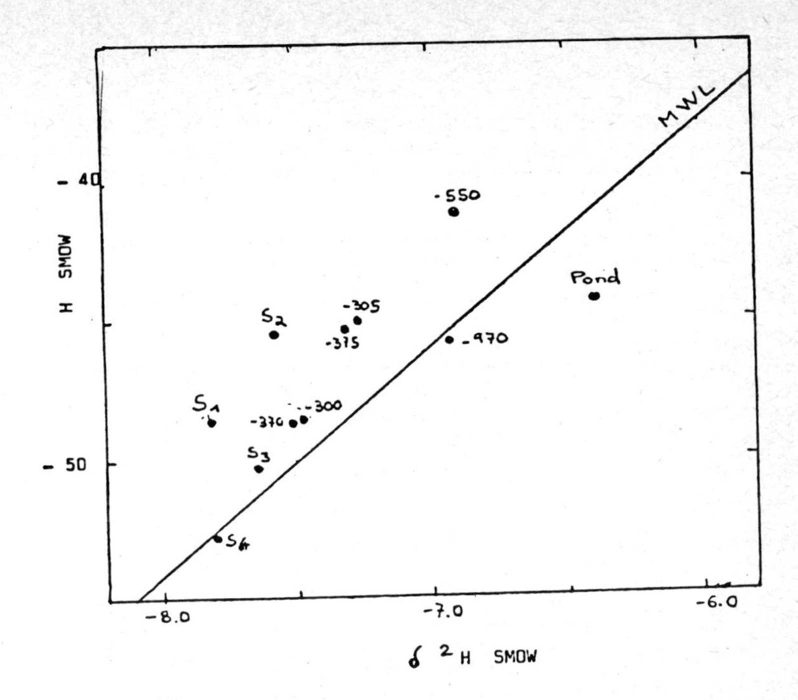

Figure 2 Oxygen 18 vs deuterium diagram

which is representative of maritime rains in this region of oceanic climate.

Pond water, below MWL, is evaporated. The composition of 300 m, 305 m, 370 m and 375 m borehole waters are close to springs waters, but heavy isotopes contents of deeper borehole waters (550 m, 970 m) are different. After drilling operations the 1000 m borehole was emptied down to 670 m. Ground-waters invaded progressively the hole. Two origins are possible for that supply : (i) intraformational deep circulation and (ii) seepage from the surficial zone where porosity and permeability are rather high in the weathered horizon. The ionic contents, the dissolved gases contents and the extremely low permeability calculated from hydraulic tests (5), are in favor of the second hypothesis. The 970 m water would be drilling (pond) water mixed with shallow groundwaters which invaded the hole after removing water until 670 m. On the ^{2}H vs ^{18}O diagram, the 970 m sample lies on a mixing line between pond water and spring water. The causes of this mixing process are molecular diffusion and/or thermal convective movement (i.e. thermal diffusion).

4.2 Tritium

Tritium contents of springs and pond water are low (29 to less than 20 T.U.). Borehole water collected at 300 m contains a part of recent water (6.3. T.U.) showing the invasion of borehole by shallow groundwaters.

5. CONCLUSIONS

Borehole waters in 1980 were probably a mixture of drilling waters and shallow groundwaters; intraformational waters corresponding to deep flows have not been evidenced.

New chemical analyses (samples 1982) show variations in ionic contents, chiefly Ca^{2} content, which could be due to either *in situ* evolution or extremely slow groundwater circulation. These results require confirmation by further sampling and measurements.

TABLE 1

ECHANTILLONS D'EAU - ANALYSES CHIMIQUES

	pH	Cond 25°C µS	Ca^{++} mg/ℓ	Mg^{++} mg/ℓ	Na^{+} mg/ℓ	K^{+} mg/ℓ	Cl^{-} mg/ℓ	SO4= mg/ℓ	NO3- mg/ℓ	$HCO3^{-}$ mg/ℓ	SiO2 mg/ℓ	Sr mg/ℓ	$NO2^{-}$ mg/ℓ	F mg/ℓ	Fe mg/ℓ	Al mg/ℓ
EAUX DE SURFACE PRELEVEES les 27 et 28/05/80																
Source haute Maison du Bois	4.4	46.7	1.35	0.93	4.21	0.23	3.0	3.0	-	12.2	8.4	0.02	-	-	-	-
Alim. eau potable - Gauche	5.6	46.3	0.85	0.37	4.63	0.39	4.25	2.0	-	7.6	10.8	0.00	-	-	-	-
Alim. eau potable - Droite	5.6	38.0	0.60	0.28	4.52	0.34	2.5	2.0	-	9.1	13.7	0.01	-	-	-	-
Source basse Maison du Bois	5.5	47.8	1.98	0.83	3.80	1.16	3.5	5.2	-	9.2	9.0	0.02	-	-	-	-
Etang	6.9	34.5	1.73	0.57	3.90	0.68	3.0	5.0	-	9.1	4.6	0.01	-	-	-	-
PHASES LIQUIDES DES ECHANTILLONS PRELEVES DANS LE FORAGE le 15/07/80																
Cote - 300 m	8.0	407	3.3	5.5	111	5.5	20	0	2	256	7.6	<0.01	<0.05	2.0	0.20	0.080
Cote - 370 m	8.1	422	3.6	1.5	103	4.55	21	0	1	250	10.0	<0.01	<0.05	2.5	0.18	0.055

TABLE 2

SITE K

ECHANTILLONS D'EAU - ANALYSES ISOTOPIQUES

	$\delta\ ^{18}O$ ‰ /SMOW ± 0.25	$\delta^{2}H$ ‰ /SMOW ± 2	^{3}H U T ± 6
EAUX DE SURFACE PRELEVEES LES 27 & 28/05/80			
Source haute Maison du Bois	- 7.73	- 47.3	29
Alim. eau potable - Gauche	- 7.58	- 45.3	< 20 (1)
Alim. eau potable - Droite	- 7.67	- 50.1	< 20 (1)
Source basse Maison du Bois	- 7.80	- 52.9	< 20 (1)
Etang	- 6.39	- 44.7	< 20 (1)
PHASES LIQUIDES DES ECHANTILLONS PRELEVES DANS LE FORAGE 1000 m le 15/07/80			
Cote - 300 m	- 7.48	- 48.5	6.3 ± 0.8(2)
Cote - 370 m	- 7.53	- 48.6	
EAUX PRELEVEES DANS LE FORAGE 1000 m le 18/07/80			
Cote - 305 m	- 7.31	- 45.2	-
Cote - 375 m	- 7.28	- 45.0	-
Cote - 550 m	- 6.91	- 41.1	-
Cote - 970 m	- 6.93	- 45.9	-

(1) inférieur à la limite de détection. Volume d'échantillon insuffisant pour un enrichissement

(2) valeur obtenue après enrichissement

REFERENCES

(1) B.R.G.M. : "Etude préliminaire du massif K". Rapport n°80 SGN 166 HYD (1980).

(2) Fontes, J.C., Michelot J.L. and Calmels P. : "Etude geochimique et isotopique des eaux du forage profond du site K" Rapport CEN-G ORIS/LABRA/SAR/LAT n° 81/13 (1981).

(3) Fritz, P., Barker, J.F. and Gale, J.E. : "Geochemistry and isotope hydrology of groundwaters in the Stripa granite". Univ. of California, Lawrence Berkeley Laboratories, Berkeley, CA, Rep. LBL - 8285 (1979).

(4) Fritz, P. and Frape, S.K. : "Saline groundwaters in the canadian shield" a first review. Chem. Geol. 36, 179-190 (1982).

(5) C.E.A. - I.P.S.N. : "Investigation par forages profonds du granite d'Auriat" Rapport C.S.D.R. 81/03 (1981).

Session 4

INVESTIGATION OF BUFFER AND BACKFILL MATERIALS

Chairman - Président

P. GNIRK

(United States)

Séance 4

RECHERCHES DANS LE DOMAINE DES MATERIAUX TAMPONS ET DE REMBLAYAGE

PRELIMINARY RESULTS FROM THE BUFFER MASS TEST OF PHASE I, STRIPA PROJECT

R. Pusch and L. Börgesson
Div of Soil Mechanics, University of Luleå
Luleå, Sweden

ABSTRACT

Major factors which determine the functions of bentonite-based buffer materials and which are investigated in the KBS 2-type test plant in Stripa, are temperature, water uptake, as well as swelling and water pressures. The recorded temperatures are well within the predicted range. The water uptake in the holes with the heaters, as interpreted from moisture gauge signals, has taken place in reasonable agreement with the predictions, which confirms that the pattern of water--bearing fractures in the rock largely determines the rate and distribution of the water uptake. The moistening of the tunnel backfill is also in fairly good agreement with the predictions although the scenario is not completely revealed yet. This is the case for the water pressure build-up as well.

1. PROCESSES OF MAJOR INTEREST

The required chemical stability of the smectite minerals depends on the temperature as well as on the ground water chemistry. A safe upper temperature limit has been taken as 100 °C of water saturated bentonite for the KBS concepts, which are in fact designed to yield temperatures not exceeding 80 °C after saturation. The actual temperatures recorded in the KBS 2-type Buffer Mass Test (BMT) are therefore of considerable interest and they will be presented and commented on in this report. The actual water uptake, which governs the swelling process and the development of swelling pressures, is also of great interest particularly because the predicted rate of moistening has been deduced only on the basis of small-scale laboratory tests and with no respect to the influence of the width of the water-bearing joints and fissures in the rock. The moistening and the associated swelling effects will therefore be major items in the report. Special attention will be paid to the water uptake in the tunnel backfill for which the predictions are relatively uncertain. The groundwater pressure build-up in the adjacent rock is also dealt with since it is partly decisive of and dependent on the water uptake in the backfill.

2. TEMPERATURE

The KBS 2 concept implies that waste-containing canisters, initially producing heat corresponding to 600 W, be surrounded by precompacted blocks of bentonite (Fig. 1). Using laboratory-derived thermal coefficients for the bentonite in its initial state (ρ=2.07-2.14 t/m^3), the surface temperature of the canisters has been estimated at 60-80 °C, the radial temperature gradient in the bentonite

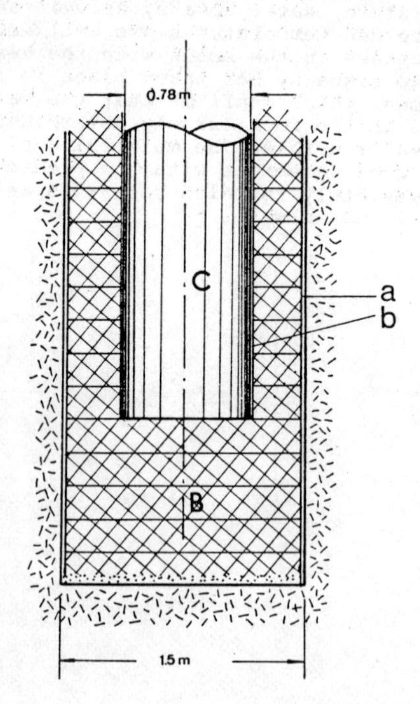

Fig. 1. The cylindrical copper canister in its clay envelope surrounded by rock. a is the outer slot, b the inner, narrow one. B is the closely fitting system of highly compacted bentonite blocks, while C is the canister.

being approximately 1 °C/cm. The geometry of the heater/bentonite/ /borehole system in Stripa was chosen so as to yield approximately the same "canister" surface temperature and temperature gradient in the bentonite.

In three of the holes (no. 1, 2 and 5), i.e. where the rock is much fractured and the groundwater inflow fairly high, the outer 10 mm wide slots (cf. Fig. 1) were left open, while the 30 mm wide slots in the remaining holes were filled with bentonite powder of low density. The different slot widths were chosen in order to obtain, ultimately, the same bulk density 2.10 t/m^3 in all the holes.

Fig. 2. illustrates a FEM-predicted location of isotherms in a cross section through holes no. 1 or 2 after 1 year [1]. The calculation does not include any correction of the heat conductivity that is expected as a consequence of the successive water uptake, neither in the compacted bentonite, nor in the tunnel backfill.

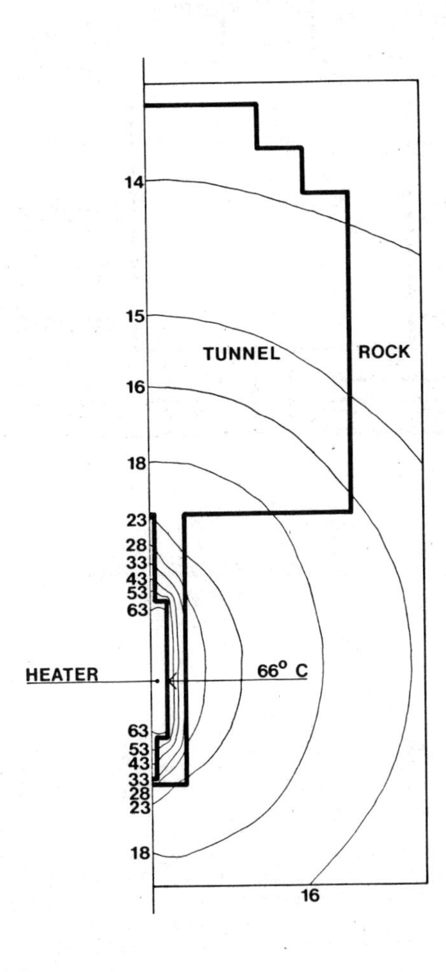

Fig. 2. Predicted distribution of temperatures in °C in cross sections through holes no. 1 and 2 after 1 year.

The main outcome of the temperature recording during the passed year is illustrated in Fig. 3. For the sake of clarity, only the temperatures after 0.2 and 1.0 years are shown, the latter being almost identical with the predicted ones. A closer look at the radial temperature distribution around the heaters shows that the water uptake in the bentonite has reduced the maximum surface temperature at mid-height of heaters no. 1 and 2 from about 70 °C after 0.2 years to 65 °C after 1 year.

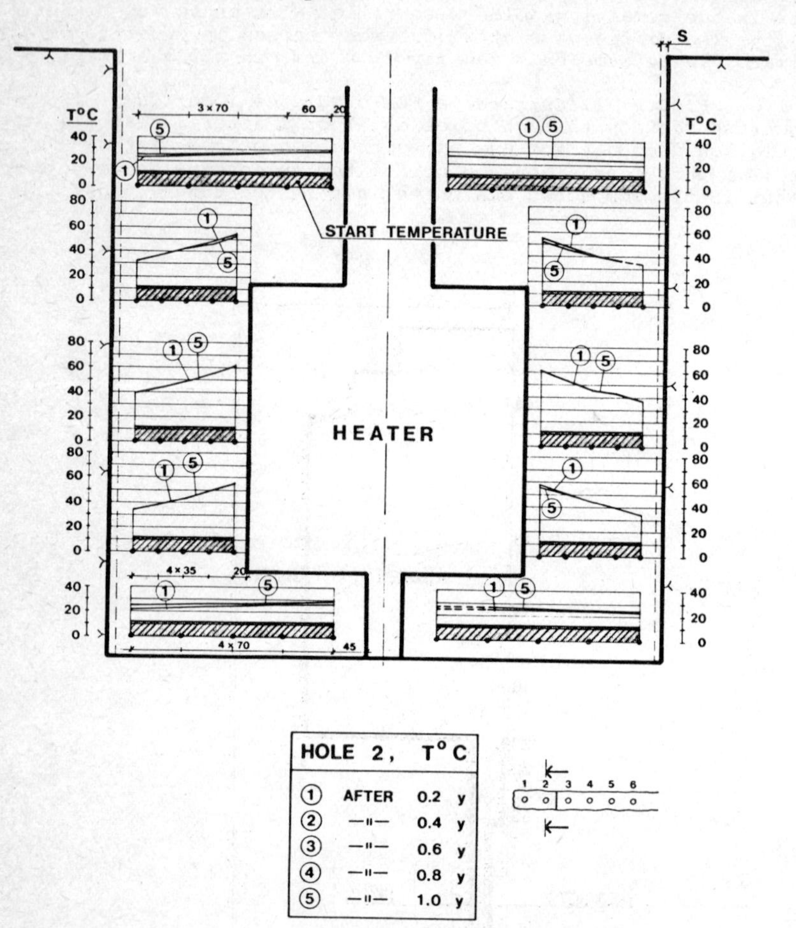

Fig. 3. Actual temperature distributions after 0.2 and 1 years, respectively, in the cross section through hole no. 2. S = original 10 mm open slot. (Different scales in horizontal and vertical directions).

The three holes with the outer slot filled with bentonite powder give higher temperatures because of the low heat conductivity of the powder in which water is accumulated non-uniformly and at a slow rate. The actual temperature distribution in the driest hole is illustrated by Fig. 4. For comparison, it should be mentioned that FEM-analyses yield a maximum heater surface temperature of about 120 °C if it assumed that the bentonite dries out[1].

[1] The condition of no water content is ficticious: "dry" corresponds to the actual water/mineral ratio obtained after heating to 105°C to which the applied λ values refer.

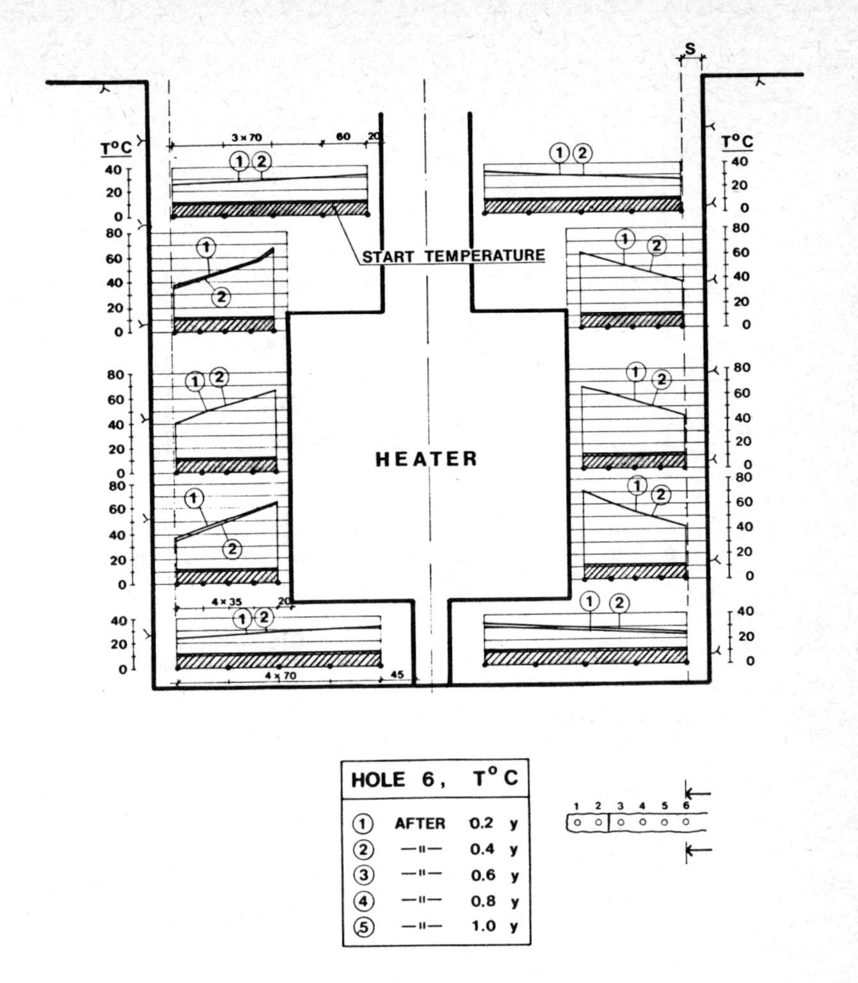

Fig. 4. Actual temperature distributions after 0.2 and 0.5 years, respectively, in the cross section through hole no. 6. S = 30 mm slot filled with bentonite powder.

3. WATER UPTAKE

3.1 Compacted bentonite

The rate and uniformity of water uptake is largely determined by the initial conditions at the bentonite/rock interface. Thus, in holes no. 1, 2 and 5, i.e. where the outer slot was left open, water filled this slot very soon after the immersion of the heater/bentonite columns in their holes. This created rapid swelling of the bentonite and the establishment of a soft, water saturated peripheral zone which then furnished the rest of the cores with water. This involved consolidation of the outer zone and considerable swelling of the core, which seems to have torn off the cable connections of a number of moisture sensors. FEM-analyses based on the assumption that water is available all over the rock/bentonite interface, and that the uptake of water can be approximated as a diffusion process where differences in water content are the driving forces, have yielded the water content distributions shown in Fig. 5.

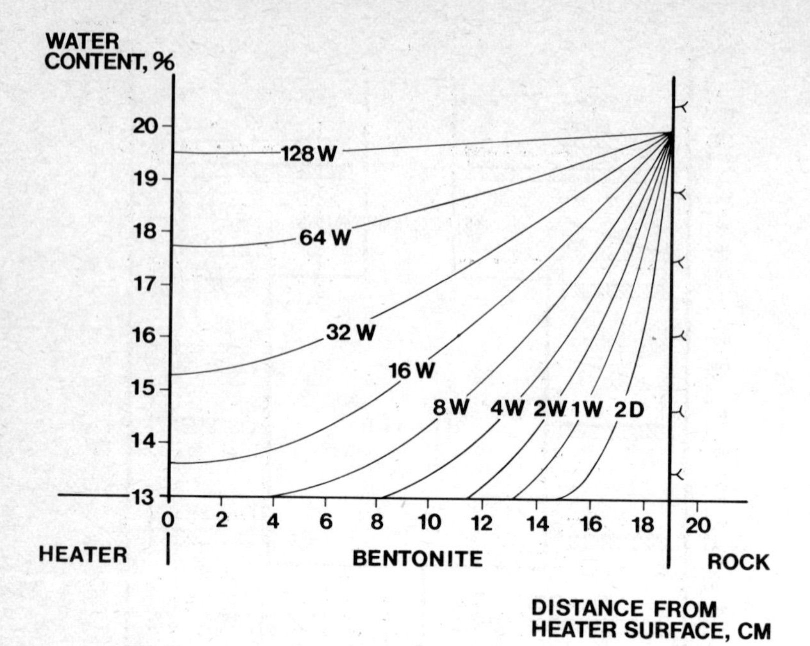

Fig. 5. Curves representing the theoretical distribution of water contents (13% being initial, 20% corresponding to complete saturation) in holes no. 1, 2 and 5. W means weeks and D days.

The predicted as well as actual conditions in hole no. 1 after 1 year are shown in Fig. 6. We see that the agreement between theory and practice is not particularly good, which is the case for holes no. 2 and 5 as well. This is in contrast to the conditions the first 6 months when the agreement was satisfactory.

In holes no. 3, 4 and 6, the bentonite powder in the outer slot absorbs water mainly from the top and bottom where the rock is fairly fractured. Additional water migrates into the powder from one or a few clearly water-bearing rock fractures, the general pattern of moistening being illustrated by the FEM-derived Fig. 7. The recording verifies the slow uptake suggested by this figure although the number of moisture sensors is far too small to reveal any details of the actual water distribution. One of these holes, namely no. 4, is the first to be opened and this is expected to yield a number of interesting features such as the possible water uptake in the bentonite also from very narrow rock fissures.

3.2 Tunnel backfill

The predicted water uptake in the backfill is very much dependent on the applied physical model for rock/backfill interaction with respect to water flow. The most probable distribution of water-bearing joints and the assumption that the water uptake can be approximated as a diffusion process, has yielded the predicted water content increase shown in Fig. 8. The moisture sensors in the backfill which are only expected to signal when the water content increases and not to yield quantitative information, give a somewhat unclear picture of the moistening and it is still too early to conclude whether the predicted pattern applies. To gain more information,

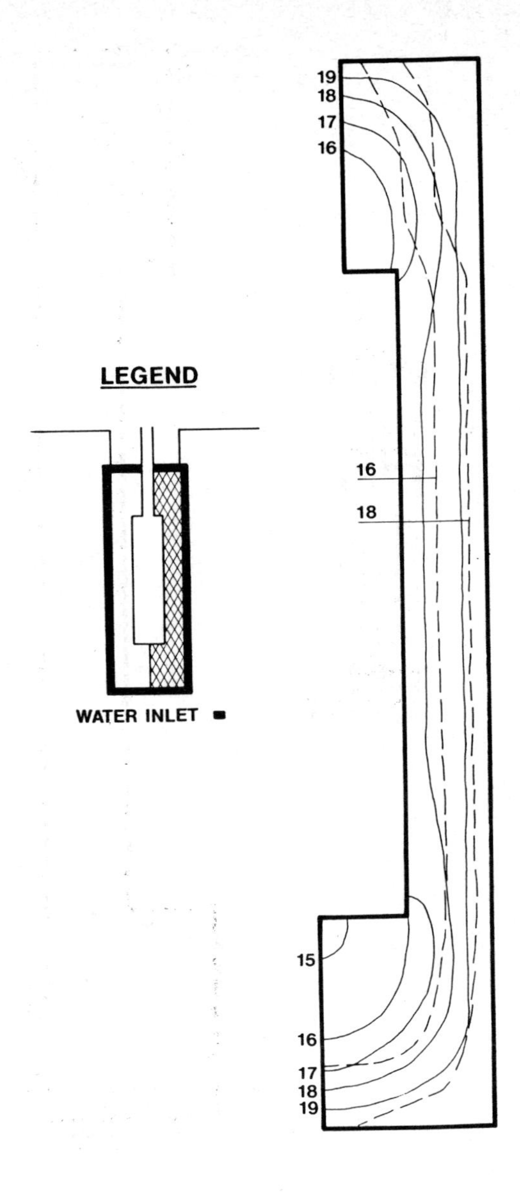

Fig. 6. Water content distribution in hole no. 1 after approximately 1 year. Broken lines represent recorded values.

continuous sampling was made in late September 1982 in two boreholes drilled through the bulwark and they both show a water content increase close to the rock surface that is similar to the graph in Fig. 8, which therefore is now taken as a basis for further investigations.

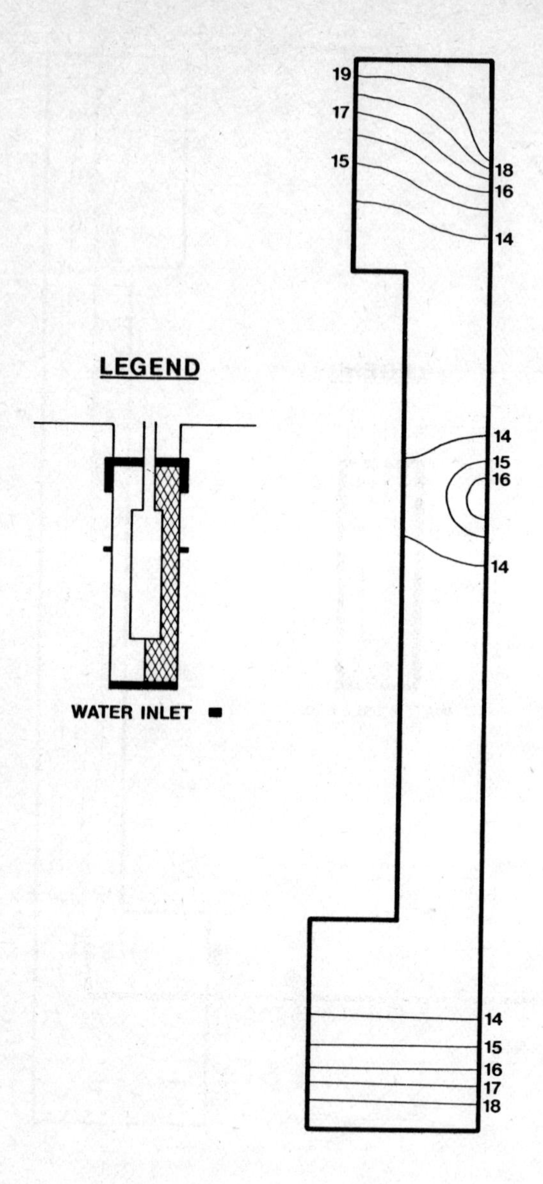

Fig. 7. Water content prediction for a hole of the type represented by no. 3, 4 and 6. The distribution of water percentages refers to 1 year after the onset of water uptake, the initial water content being taken as 13%. No reference to the influence of bentonite powder in the outer slot has been made.

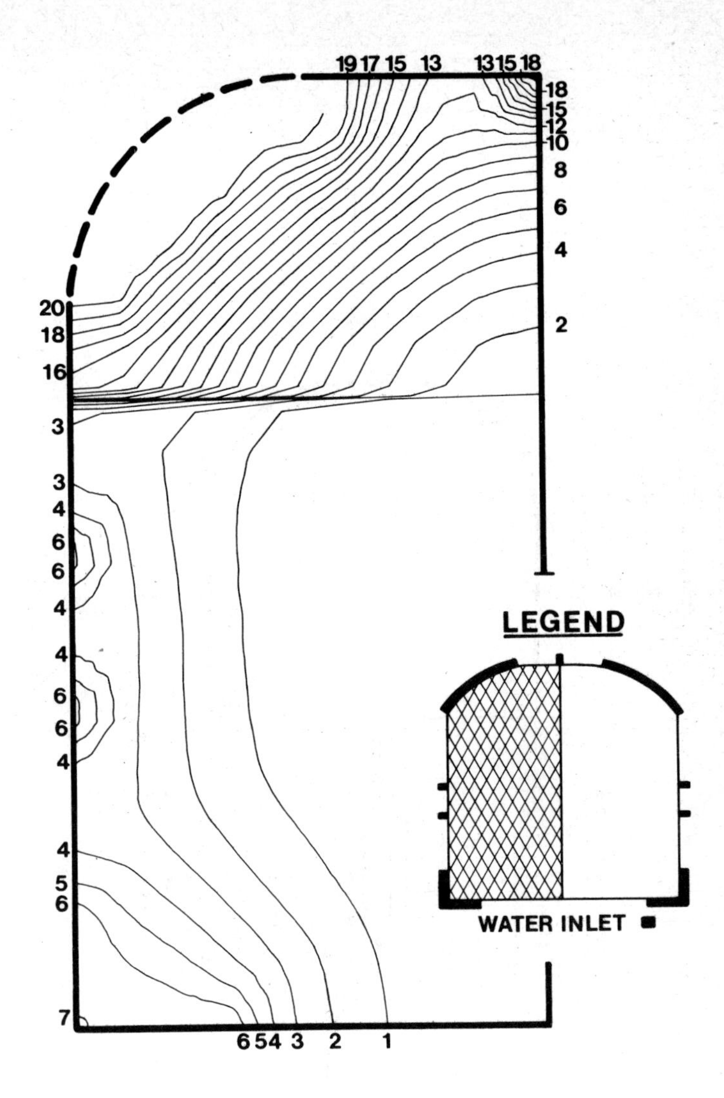

Fig. 8. Predicted increase in water content (percent units) in tunnel cross sections through holes no. 1 and 2 after 1 year. Height of tunnel is about 4.5 m.

4. SWELLING AND WATER PRESSURES

4.1 Heater holes

As expected, the swelling pressure in the heater holes is primarily determined by the degree of water saturation. Thus, the highest pressures are recorded in the "wet" holes no. 1 and 2, the maximum being about 6.4 MPa at mid-height of heater no. 2 in late September. Naturally, the successive pressure build-up, which is illustrated by Fig. 9 for hole no. 2, is much slower in the "dry" holes no. 3, 4 and 6 than in the wet ones. It is remarkable, however, that the pressures in the latter holes are significant although the moisture sensors in the bentonite core have hardly reacted yet. This suggests that water is primarily distributed in the powder of the outer slot. The water pressures are usually low compared to the swelling pressures, the highest ones being 700 kPa in hole no. 1, 620 kPa in hole no. 4, and 430 kPa in hole no. 2.

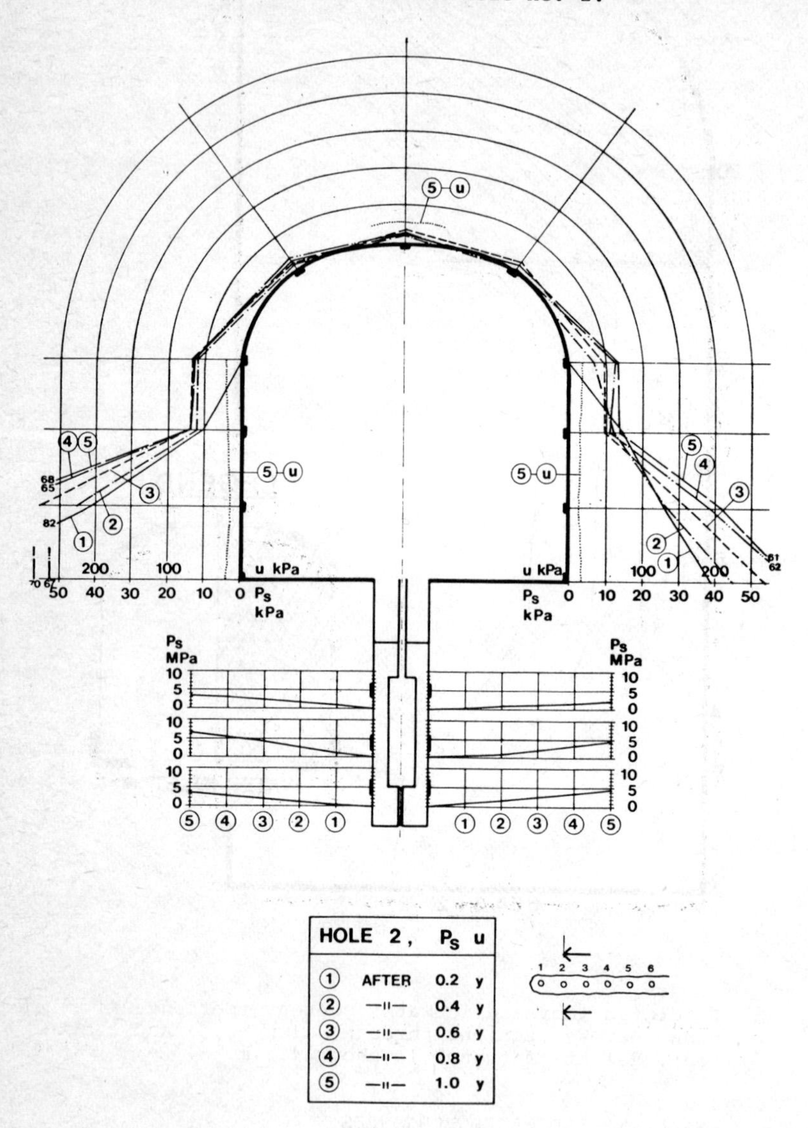

Fig. 9. Recorded pressure build-up in the cross section through hole no. 2. p_s = swelling pressure; u = water pressure.

4.2 Tunnel

The pressure distribution as recorded by use of Gloetzl cells in the backfill is very complex since it is induced not only by water uptake and external water pressure but also by temperature effects, by the own weight of the fill, and by swelling pressures transferred from the highly compacted bentonite in the heater holes. We will confine ourselves here to comment on the pressures recorded at the rock/backfill interface and on the water pressures close to this interface. These pressures are plotted in Fig. 9 for the cross section through hole no. 2, and it is concluded from this diagram, as well as from the water content determination in the two recent boreholes, that water pressures – although low so far – are now being built up around the larger part of the tunnel periphery. The Gloetzl cell pressures in the lower part of the tunnel walls are consistently higher, by 20 to 60 kPa, than the water pressures indicating that swelling pressures seem now to be developed.

5. REFERENCE

[1] Börgesson, L.: "Buffer Mass Test – Predictions of the Behaviour of the Bentonite-based Buffer Materials". Stripa Project Internal Report 82-08.

Session 5

PROGRAMMES FOR FURTHER IN SITU EXPERIMENTAL WORK

Chairman - Président

L.-B. NILSSON

(KBS)

Séance 5

PROGRAMMES FUTURS DES TRAVAUX EXPERIMENTAUX IN-SITU

THE PROPOSED PHASE II OF THE INTERNATIONAL STRIPA PROJECT

compiled by
Hans S Carlsson
Swedish Nuclear Fuel Supply Co/Division KBS
Stockholm, Sweden

ABSTRACT

A brief description of the research program for the accomplishment of the proposed second phase of the Stripa Project is given. The tentative program, scheduled to be started in early 1983 and completed in 1986, includes technical proposals that cover the following investigations:

1) Detection and characterization of fracture zones in crystalline rocks by using cross-hole geophysical and hydraulic methods; 2) tracer experiments in fractured granite; 3) diffusion experiments in highly compacted bentonite and bentonite/sand mixture and finally 4) sealing of boreholes and shafts using highly compacted bentonite.

1 BACKGROUND

The ongoing Stripa Project is scheduled to be completed in May 1984. A proposed program for a second phase of the project, scheduled to be started in January 1983 and completed at the end of 1986, is currently being considered by the potential member countries. The proposed program consists of research in the following areas:

- detection and characterization of fracture zones by using crosshole geophysical and hydraulic methods
- tracer experiments in fractured granite
- diffusion experiments in highly compacted bentonite and bentonite/sand mixture
- sealing of boreholes and shafts using highly compacted bentonite

Given below is a brief description of the program. The estimated cost, including an annual inflation of 10% is MSEK 63.0 for the four-year period. An interest to participate in the possible second phase of the Stripa Project has been declared by Canada, Finland, France, Japan, Sweden, Switzerland, United Kingdom and the United States.

2 CROSS-HOLE TECHNIQUES FOR THE DETECTION AND CHARACTERIZATION OF FRACTURE ZONES IN THE VICINITY OF A REPOSITORY

2.1. Objective

The current proposal /1/ has been set up in cooperation between the Geological Survey of Sweden (SGU), the Swedish National Research Defense Institute, and the Institute of Geological Sciences, United Kingdom. The principal investigator is Dr Olle Olsson, SGU.

The purpose of the program is to develop techniques and instrumentation with the capability to detect fracture zones in a specified volume of investigation in crystalline bedrock. The methods developed should have the capability to determine the location, extent, and thickness of fracture zones and also give a quantitative measure of the bedrock quality and its influence on the storage of radioactive wastes. The program also intends to resolve questions concerning the possible resolution and the uniqueness of the models of bedrock obtained with these methods.

In designing equipment for this type of techniques it is important to obtain optimal resolution. The resolution is a function of the range of volume of rock that will be investigated and it is therefore meaningful to separate the development of these techniques into a large scale and a small scale investigation.

2.1.1 Large scale investigation

With a high degree of confidence find the location, extent, and properties of major fracture zones in crystalline rocks with a thickness of the order of 10 m, in a volume reaching about 500 m from the positions accessible by instrumentation, such as boreholes, drifts, and the ground surface. This definition will include the major high permeability zones which may facilitate rapid transport of the small amount of radioactive material that might enter the zones. The presence of zones of this magnitude will also influence the design and construction work.

2.1.2 Small scale investigation

With a high degree of accuracy find the location, extent, and properties of fracture zones, with a thickness of the order of 0.1 m, in a volume reaching about 50 m from the positions accessible by instrumentation. In this case instrumentation is to be placed in drifts or in boreholes made from drifts. This definition will include the fractures with intermediate permeability.

2.2 Methods to be used

Considering the definition of the purpose given above and a discussion about methods which may be used /2/ it is proposed that the main effort should be devoted to the development of instrumentation, measurement procedures and interpretation techniques for the cross-hole seismic method in the large scale investigation and the high frequency electromagnetic technique (radar) in the small scale investigation. However, the cross-hole seismic method as well as the correlation with relevant hydraulic properties of the rock is also proposed to be included in the small scale investigation. This combined approach will make it possible to develop a set of methods which will have the capability to adequately locate and describe the major features of the bedrock in different scales. Three dimensional models will be designed where the bedrock is considered to consist of parts of homogeneous bedrock separated by fracture planes.

2.3 Sites to be used

The small scale site with dimensions of about 100 meters containing one or more minor water yielding fracture zones will be most suitable for the borehole radar and the detailed hydraulic wave program, but will also be used for the seismic program. These requirements would be met by the fracture zone intersected at 130 m in the borehole E1 at the 360 meter level in the Stripa Mine, see Figure 1. At this site additional drilling is required to provide for appropriate research conditions.

The large scale site, most favorable for the cross-hole seismic program, requires distances between boreholes of several hundred meters. For this purpose the Swedish test site Gideå is judged more suitable than the Stripa Mine. At this site a comprehensive site investigation program has been conducted. The investigations include surface geological mapping, detailed geophysical measurements with magnetic, electromagnetic, electric and seismic techniques on an area of 5 km^2. To investigate the properties of fracture zones at depth 10 diamond boreholes, each about 700 m long, and about 25 percussion boreholes, each about 100 m long, have been drilled. In the diamond boreholes detailed core log and single hole geophysical and hydraulic measurements have been made. This area is probably one of the most thoroughly investigated sites at present and the existing knowledge of fracture zones makes it an ideal site for testing of new methods.

3 THREE DIMENSIONAL TRACER EXPERIMENT

3.1 Objective

Tracer experiments reported to date will not supply any data which can be utilized to understand the spacial distribution of pathways. Nor will they give any indication of tracer movement normal to the average flow direction. There are no experimental results available in the literature or other sources known to us on the spacial distribution of flow in fissured crystalline rock at large depths (<150 m).

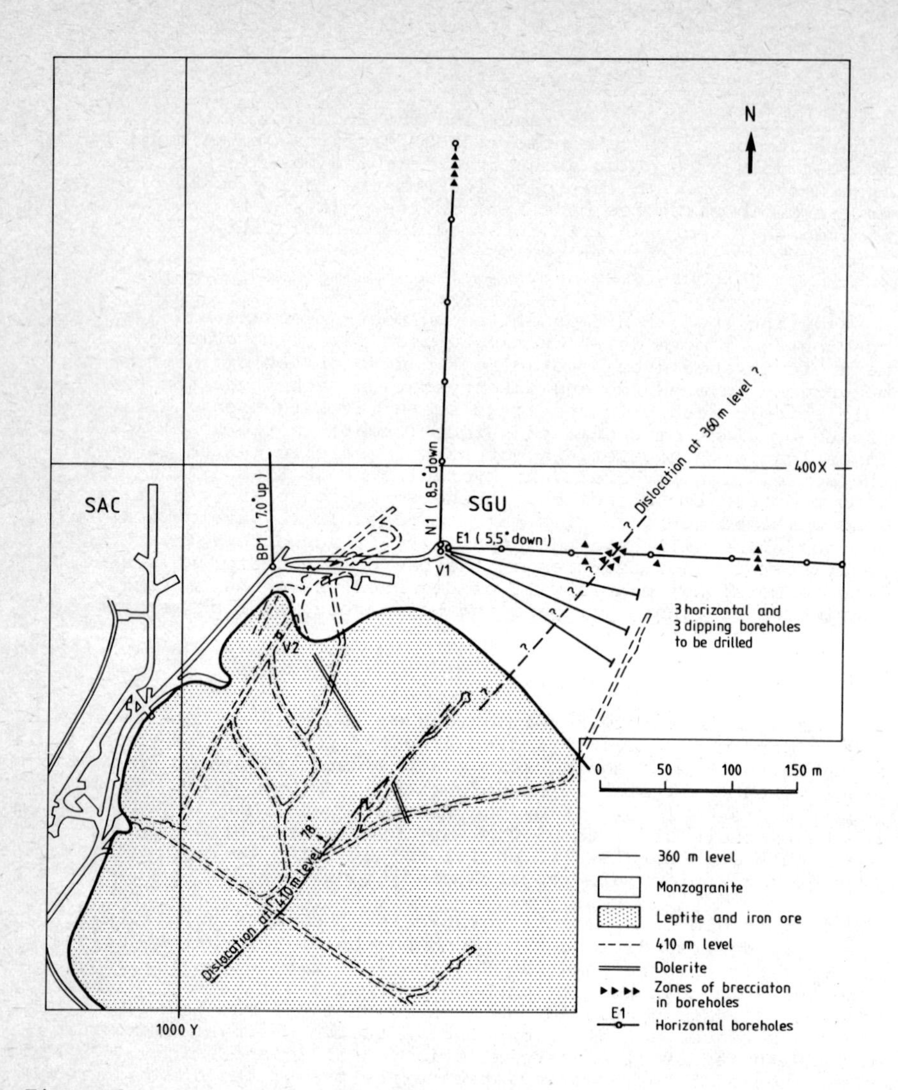

Figure 1. Plan view of the Stripa Mine at the 360 m level

As channeling effects and transverse dispersion can have a profound influence on the arrival times and concentrations of radionuclides escaping from a repository, it is deemed important that these effects should be studied.

The purpose of the proposed study is thus

- to develop techniques for large scale tracer experiments in low permeability fissured rock
- to determine flow porosity
- to study longitudinal and transverse dispersion in fissured rock
- to study channeling
- to obtain data for model verification and/or modification

The investigation given below /3/ is proposed by Professor Ivars Neretnieks, Royal Institute of Technology, Stockholm, Sweden.

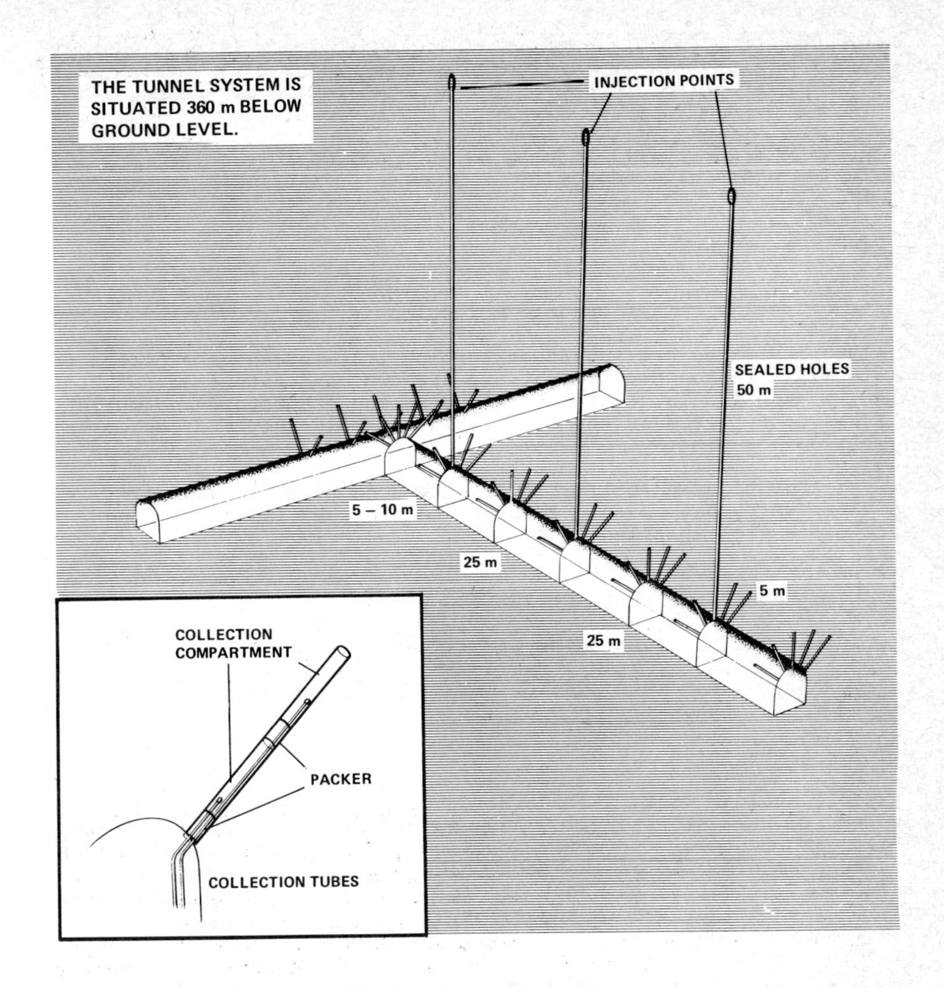

Figure 2. Schematic layout of the 3-D migration experiment

3.2 Proposed investigation

Figure 2 schematically shows the arrangement of injection points and collection holes around a tunnel system. A set of non-sorbing tracers are injected simultaneously in the far end and at intermediate points of the otherwise sealed injection holes. Different tracers are used in different injection points. As the tunnel is well below the water table, the tracers will flow towards the tunnel. The arrival of the tracers to the tunnel is monitored in the collection holes near the tunnel. The collection holes are arranged in such a way that they cover a large volume of the rock. They are placed where they intersect larger "wet" fissures. The tracer from each injection point can then be monitored in a large number of holes arranged in such a way that the transverse spreading of the tracer can be measured. By injecting different tracers in the different injection holes, the intersection of flow paths can also be studied. The tracers from the hole nearest the tunnel intersection can be monitored in both tunnels.

It is expected that at least 3-4 injection holes will be drilled and that at least 40 collection holes are necessary. The length of the injection holes are about 50 m and the length of the collection holes are 5-10 m.

Non-active tracers will be used as much as possible. The non-sorbing tracers will be:

Uranine
Br^-
I^-
Cr - EDTA
Large molecular weight organic molecules

If the flow paths are such that sorbing tracers can be expected to arrive in a reasonable time, there are two candidate tracers which have been tested in natural fissures in the laboratory and which will have been used in the fissure in Stripa. They are Cesium and Strontium.

Tracer injection and collection can be done by techniques similar to those which have been developed and are used in the ongoing single fissure experiment.

4 BOREHOLE AND SHAFT SEALING TESTS

The investigation given below /4/ is set up in cooperation between University of Luleå, (LuH), Sweden and the Swedish Nuclear Fuel Supply Co/Division KBS. The principal investigator is Professor Roland Pusch, LuH.

4.1 Objective

Terminal nuclear waste storage at great depths requires plug systems to prevent water flow and ion migration through boreholes, shafts and tunnels. Effective sealing requires that the plug is at least as impermeable as the rock which is replaced by the plug. This implies that no passages are formed along the rock/plug interface. These requirements suggest the use of dense, smectite-rich clays such as bentonite for plugging purposes. A possible technique has been suggested for borehole plugging using highly compacted bentonite in perforated copper tubes /5/. A series of laboratory tests and a field study indicate that the technique should be applicable on a full scale.

The proposal is divided into three parts which are described below.

4.2 Part I - Borehole Plugging

This part starts with a pilot test using a Ø 76 mm hole to demonstrate the general principle and the technique for in situ determination of the sealing function of the plug. The site will be close to an existing Ø 1 m vertical shaft extending from the roof of one of the drifts in the mine. The hole will be used also to yield basic rock information for a suggested shaft plugging test (see Part II). In addition, an existing 100 m long Ø 56 mm horizontal borehole will be plugged and tested.

4.3 Part II - Shaft Plugging

A 10 m long vertical shaft made partly by slot drilling technique and partly by blasting extends between two drifts in the mine. This shaft is intended for a test in which the effect of a clay plug on the paths and rate of flow of groundwater in the adjacent rock will be investigated.

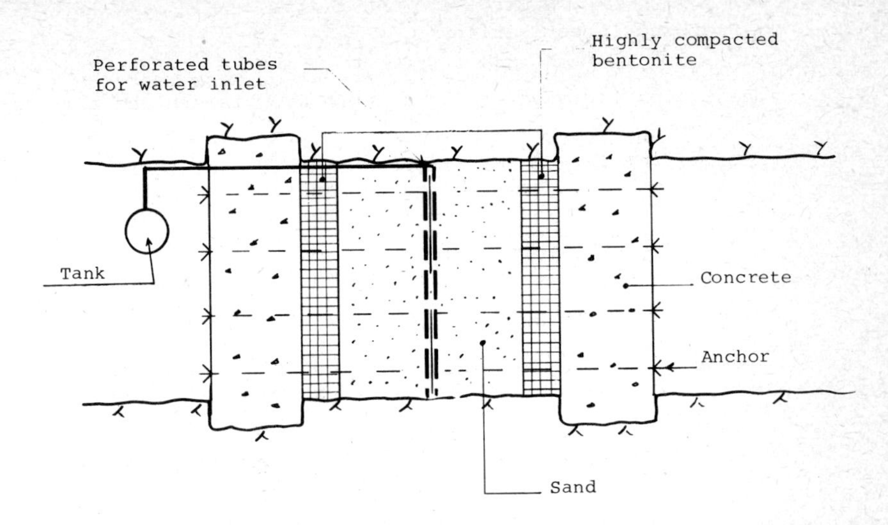

Figure 3. Schematic experimental layout of the test regarding the isolation of a highly water bearing zone

4.4 PART III - Isolation of a drift from a richly water-bearing rock zone by use of highly compacted bentonite

The project concerns isolation of a richly water-bearing zone which crosses a drift, tunnel or shaft. The test conditions are artificial in the sense that this zone is created by applying a series of perforated pipes around the periphery of a tunnel, which will be excavated in rock poor in fractures. The pipes are embedded in a sand backfill applied between two concrete bulkheads which support the sealings consisting of highly compacted bentonite (Figure 3). Water is injected at a pressure up to 300 m waterhead into the central sand zone, which simulates crushed rock, and the pressure and leakage are measured. The swelling pressure is measured separately.

5 DIFFUSION OF NUCLIDES IN A REPOSITORY

The following investigation is proposed by Dr Olli Heinonen, Technical Research Centre of Finland.

5.1 Objective

The proposed investigation has the following main objectives:

- To observe the migration of sorbing and non-sorbing tracers under controlled and defined conditions in a repository with different back-fill materials;
- To perform parallel small and intermediate scale laboratory experiments to obtain a basis for comparison of field and laboratory data;
- To fit the experimental results to those of different mass transport models to get quantitative values for diffusion coefficients of tracer-nuclides in different back-fill materials;
- To develop feasible techniques for investigating migration phenomena in a real repository environment.

5.2 Proposed Investigation

The field tests will be performed in the Buffer Mass Test (BMT) site of the Stripa I project to obtain proper moisture and temperature conditions.

The tracers will be introduced into the compacted bentonite of deposition hole No. 5 between the heater and the bentonite as well as between the bentonite and surrounding rock. Two other injections will be made into the tunnel filled by 10%/90% bentonite/quartz sand mixture. After approximately two years samples will be excavated for concentration profile studies in the laboratory.

Since the tracers will migrate only in the wet bentonite the injection will start when the data gathered during the BMT will show favorable conditions for the experiments. The length of the experiment will be adjusted on the basis of results of laboratory studies to obtain maximum benefit of the site investigations.

The water chemistry has a strong influence on the migration of radionuclides. The physicochemical conditions will be analyzed during the experiment by in situ measurements and by sampling from deposition hole No. 4 and from post experimental excavated bentonite samples.

5.3 Tracers to be used

The diffusion of nuclides in bentonite is measured by means of the following tracers, which cover the most important chemical species of the wastes: Tritium, Cs, I, Ni, Eh, Th and U.

The tracer experiments are performed in the field and in the laboratory. In addition to the above mentioned, stable tracers for certain nuclides, e.g. Tc, Np and Pu, experiments with radionuclides are made in the laboratory.

6 TRACER TESTS PROPOSED BY UKAEA

The following investigation /6/ is proposed by Dr P Bourke, Atomic Energy Research Establishment, United Kingdom.

6.1 General

Various hydrogeological investigations, including the KBS-LBL work at Stripa, have shown that leakage of radionuclides from waste deeply buried in hard rock back to the surface will be mainly by transport in water flow through discrete fractures.

The following alternative, combined experimental and theoretical approach to the fracture hydrology, is therefore being developed. It is assumed that the fractures forming the pattern through which most of the water flows, are planar and have the following variables:

- frequency of occurrence or number per unit volume of rock
- orientation
- lengths between ends of characteristic linear dimensions
- effective hydraulic aperture

Theoretical work is being done at Harwell /7/ and other establishments to develop percolation models which, given experimental statistics for the means and distributions of the above variables, should be able to predict the flows.

The reason for this approach to the work is that the retardation of the radionuclide movement depends on the rates of diffusion into the rock between fractures and on the thickness of rock between the fractures.

Experimental methods have been developed in Cornwall to obtain these statistics at 200 m depths in granite. Fracture occurrence determined from multi-packer tests in single holes /8/. Radioactive tracers are then pumped into single fractures and detectors are repeatedly lowered and raised in adjacent holes to locate the positions of single or multiple arrivals of tracers - these results provide orientation and topological data about the fractures and their interconnections /9/. Finally pairs of double packers in adjacent holes are set at positions shown by the tracers to be on the same fractures. Interhole flow and pressure drop measurements are made to obtain estimates of their effective hydraulic apertures.

6.2 Objective

This proposal is simply that the above methods should be tried underground at Stripa to obtain fracture statistics for another granite. It has the merit that because these methods are proven, there is a reasonable hope that useful results can be obtained fairly quickly with moderate expenditure.

6.3 Proposed investigation

Three parallel, 56 mm diameter, cored holes forming an isosceles or right angled triangel with base and height of about 5 and 10 m respectively are required. The start and direction of these holes could conveniently be in any tunnel at 300 to 400 m depth and vertically downward. The length of the boreholes should be around 150 m.

Double packers with a 1 m separation should be used to test every meter of the lowest 100 m of one hole and to locate the major fractures intersected it. If the fracture pattern is similar to that in Cornwall some ten fractures will be found.

The packers should then be again used to pump radioactive bromine and/or iodine tracers into each fracture separately while detectors are used to locate tracer arrivals in the other holes. Both single and multiple arrivals may be found. Finally pressure drop measurements between holes should be made for flow through single fractures.

Logging of the core from the drill holes should also be carefully done to identify, using the hydraulic data from the holes, those fractures in the core which take water.

7 REFERENCES

/1/ Olsson, O., Black, J., Carlsson, L., Israelsson, H.: "Cross-hole techniques for the detection and characterization of fracture zones in the vicinity of a repository", Stripa Project, Phase II proposal, Stockholm 1982.

/2/ Olsson, O., Andersson, J.-E.: "Cross-hole geophysical methods for the detection and characterization of fracture zones in the vicinity of a repository", Stripa Project, State of the art report, Uppsala 1981.

/3/ Neretnieks, I.: "A three-dimensional tracer experiment in Stripa", Stripa Project, Phase II proposal, Stockholm 1982.

/4/ Pusch, R., Bergström, A.: "Borehole and shaft plugging tests in Stripa", Stripa Project, Phase II proposal, Stockholm 1982.

/5/ Pusch, R.: "Borehole sealing with highly compacted Na bentonite", KBS Technical Report 81-09, Stockholm 1981.

/6/ Bourke, P.J.: "UKAEA proposal for participation in the fracture hydrology research of the second stage of the Stripa Project", Stripa Project, Phase II proposal, Stockholm 1982.

/7/ Bourke, P.J., Gale, J.E., Hodgkinson, D.P., Witherspoon, P.A.: "Proc. of NEA workshop on low permeability rock", pp. 157-171, Paris 1979.

/8/ Bourke, P.J., Bromley, A.V., Rae, J., Sincock, K.: "Proc. of NEA workshop on siting repositories", pp. 173-190, Paris 1981.

/9/ Bourke, P.J., Evans, G.V., Hodgkinson, D.P., Ivanovich, M.: "AERE Report 10046".

THE RESEARCH PROGRAM AT THE CANADIAN UNDERGROUND RESEARCH LABORATORY

C.C. Davison and G.R. Simmons
Whiteshell Nuclear Research Establishment
Atomic Energy of Canada Limited
Pinawa, Manitoba ROE 1LO

ABSTRACT

Atomic Energy of Canada Limited is planning an underground research laboratory (URL) in an undisturbed portion of a granitic intrusive, the Lac du Bonnet batholith. The URL is being used to assess and improve our ability to interpret and predict geologic, geochemical and hydrogeologic conditions in large volumes of plutonic rock. As well, the URL will provide an opportunity to study the effects of excavation on the rock mass, to develop and assess shaft and drift seals, and to assess the accuracy of many mathematical models used to predict the near-field response of complex multicomponent systems. The Site Evaluation and Underground Experimental Programs, and the design[1] and construction process are discussed.

1 Footnote Added in Proof
The design discussed in this paper has two levels at approximate depths of 130 and 240 metres. The design concept presently favoured (1983 January) has a single test level with test rooms at depths between 210 and 240 metres. This new concept will be discussed elsewhere.

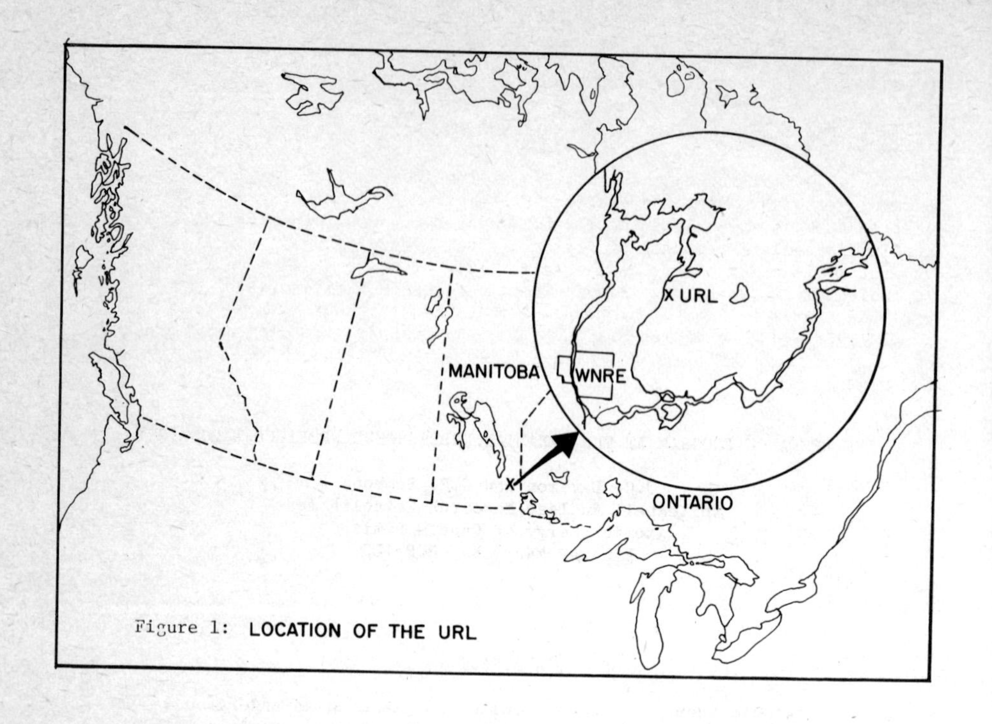

Figure 1: LOCATION OF THE URL

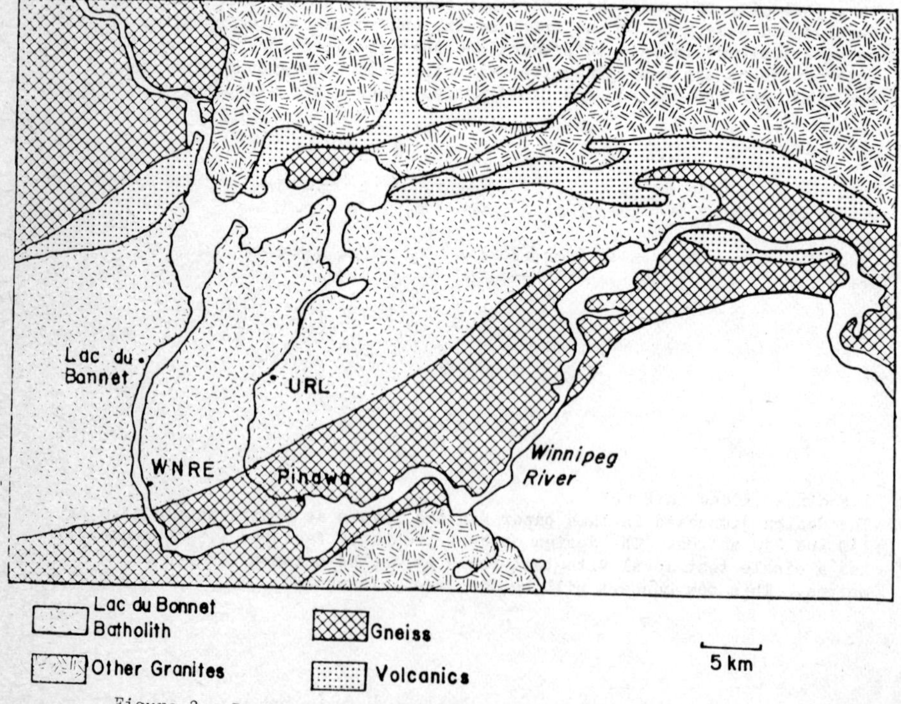

Figure 2: Regional Geologic Setting of URL and WNRE Sites

INTRODUCTION

One of the major objectives of the geotechnical research being carried out within Canada's Nuclear Fuel Waste Management Program is to develop methods of determining the geologic, geophysical and hydrogeological conditions of large volumes of plutonic rock. These conditions must be understood in order to assess the suitability of the rock mass as a barrier to radionuclide migration. Atomic Energy of Canada Limited (AECL) has employed a wide variety of airborne, surface and borehole methods to investigate plutonic rocks, and many of these methods have been applied at small-scale research sites at several different locations on the Canadian Precambrian Shield.

In 1980, AECL leased 3.8 km^2 of land from the government of the province of Manitoba, Canada with the intention of constructing an underground research laboratory (URL). The main geotechnical objectives of the URL are to assess and improve our ability to interpret and predict the geologic, geochemical and hydrogeologic conditions of large volumes of plutonic rock, to assess the effects of excavation on the rock mass, to develop shaft and drift seals, and to assess the accuracy of the mathematical models used to predict near-field conditions in the rock mass [1]. Subsurface conditions within the granitic rock mass at the URL site will be investigated prior to construction of the shaft and underground workings. Based on these measurements, predictions will be made of the geochemical and hydrogeologic disturbance that will be created in the area as construction of the shaft and underground workings proceeds. Actual conditions in the rock mass will be monitored during the development of the URL and these will be compared to the predictions. In this manner, computer models of the hydrogeology of the fractured rock system can be developed and validated.

The URL site is located about 12 km east of the town of Lac du Bonnet, Manitoba, Canada (Figure 1). This area was chosen because:

(1) It is an undisturbed plutonic rock site located well within the boundaries of the granitic Lac du Bonnet batholith (Figure 2), which is, in many ways, representative of the granitic intrusives found on the Canadian Precambrian Shield.

(2) The bedrock exposure (over 70% rock surface exposed) permits surface geological and geophysical methods of investigation.

(3) The area is located close to the service and support facilities of AECL's Whiteshell Nuclear Research Establishment (WNRE) (Figure 3).

This paper outlines the various geotechnical research aspects of Canada's URL project, including the preconstruction evaluation of the geologic, geophysical and hydrogeologic conditions of the URL site, which is currently in progress, and the underground experimental program, which is planned to begin with shaft construction early in 1984.

SITE EVALUATION PROGRAM

Since 1980, a comprehensive program of geological, geophysical and hydrogeological investigations has been underway at the URL lease area. The initial objective of the site evaluation program was to select a site for the shaft and underground facilities of the URL that would maximize the options for future experiments both in the underground facility and in the remaining lease area. In addition, preconstruction geologic and hydrogeologic conditions are being defined at the site in order to predict the perturbation that will be created as the URL shaft and underground workings are constructed. A detailed network of instrumentation is being placed in the rock mass surrounding the URL excavation to monitor the actual

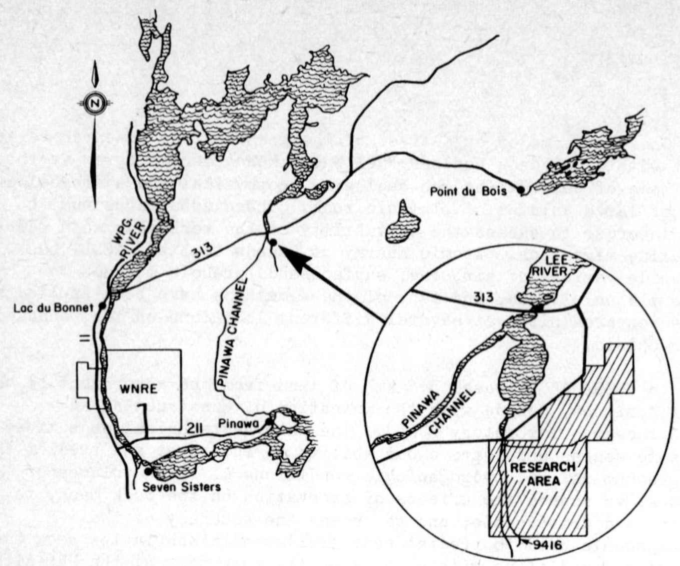

Figure 3: Location of the URL Lease Area

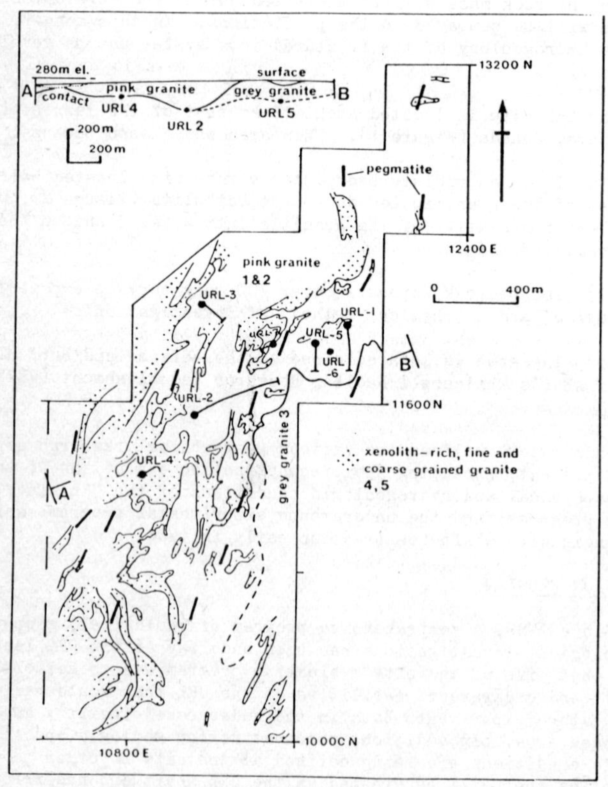

Figure 4: Distribution of Lithologic Units at Surface of URL Site

perturbation effects. These data will be used to assess the accuracy of the predictions.

Geological and Geophysical Investigations

Detailed geological and geophysical surveys have been conducted at the URL lease area by researchers from AECL and the Department of Energy, Mines and Resources (EMR) [2]. Surface mapping of the lithology at outcrop areas indicated the presence of several phases of granite. Most of the outcrop area is pink-coloured porphyritic granite with biotite-rich, biotite-poor, gneissic-foliated and zenolith-rich phases. An older, grey-coloured rich hornblende-biotite granite is located along the east margin of the lease area. Figure 4 shows the distribution of these phases at ground surface. Detailed mapping of fractures that are visible at bedrock outcroppings has indicated that the pink-coloured porphyritic granite is moderately to highly fractured, whereas the grey-coloured hornblende porphyritic granite is virtually unfractured. The dominant fracture orientation is NNE. Surface resistivity surveys compare well with the surface fracture density map, indicating that resistivity surveys can be used as a rapid reconnaissance tool to provide useful information regarding the intensity of surface fracturing in outcrop areas.

Analysis of geologic, geophysical and hydrogeologic survey data of the surface of the URL lease area indicated that the northeast portion of the lease area was the prime area of interest for the possible location of the URL facility (Figure 4). Test drilling was initiated in this area during 1981 February to provide three-dimensional subsurface information regarding the contact between the fractured pink-coloured granite phase and the unfractured grey-coloured granite phase. The initial objective of the exploratory drilling was to search for a region of the subsurface within the lease area in which a single working level of the URL underground facility could be constructed at a depth of 250 m to 350 m to allow experimental access to both the unfractured grey granite and the fractured pink granite.

From 1981 mid-February to 1981 September, boreholes URL-1 through URL-4 were drilled, using NQ continuous wireline coring methods, to probe the contact between the pink and grey granite phases. The data obtained from core logging and borehole surveys indicated that the contact between the two granite phases did not continue to dip away from the surface exposure of the grey granite. The contact was encountered at a depth of 90 m to 150 m in all of these boreholes.

Borehole URL-5 was drilled 180 m to the west of URL-1 during 1981 September to determine the extent and orientation of a large fracture zone that was intersected at a depth of approximately 325 m in URl-1. URL-5 penetrated the same fracture zone at a depth of 256 m, which indicated that the fracture zone was dipping gently (~20°) to the west. Subsequent geologic information indicates a NNE strike of 320° for this fracture zone.

In order to maximize the options available for underground experiments, given the site conditions as determined from the initial test drilling program, it was recommended that the URL should be designed with two underground levels. An upper level, at a depth of 100 m to 150 m, would allow access to the highly fractured pink granite to enable large-scale fracture-related experiments to be performed. A lower level, at a depth of 250 m to 350 m, would allow access to the unfractured grey granite. It was also considered valuable to try to provide experimental access to the subhorizontal deep fracture zone encountered in boreholes URL-1 and URL-5. Hydrofracturing results from URL-1 indicated that there may be a major stress concentration in the rock mass immediately above the deep fracture zone. By locating test rooms in this high-stress environment, conditions representative of much greater depth can be obtained in a relatively shallow excavation.

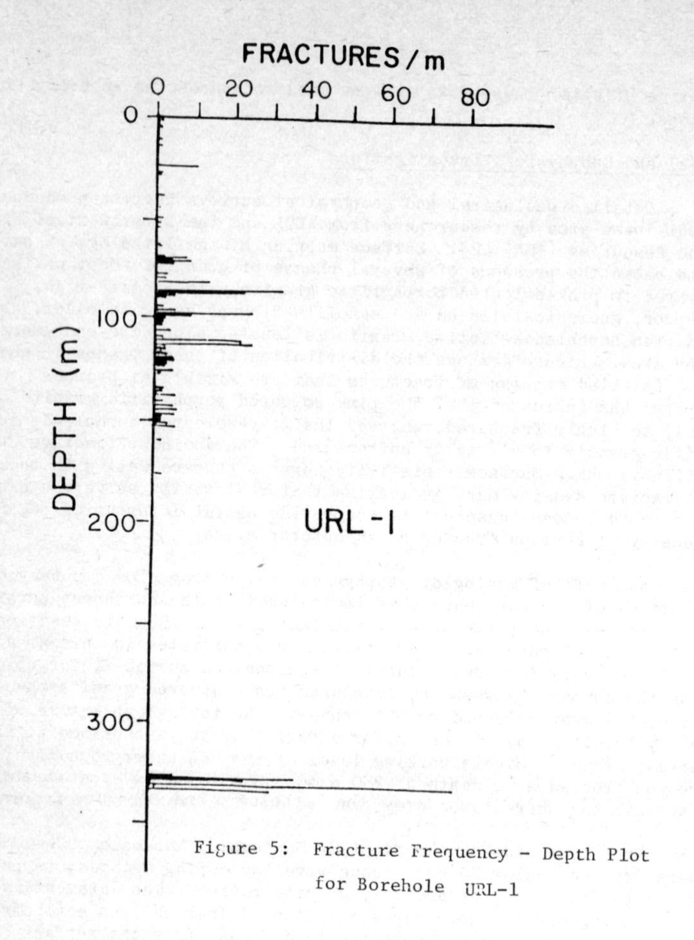

Figure 5: Fracture Frequency - Depth Plot for Borehole URL-1

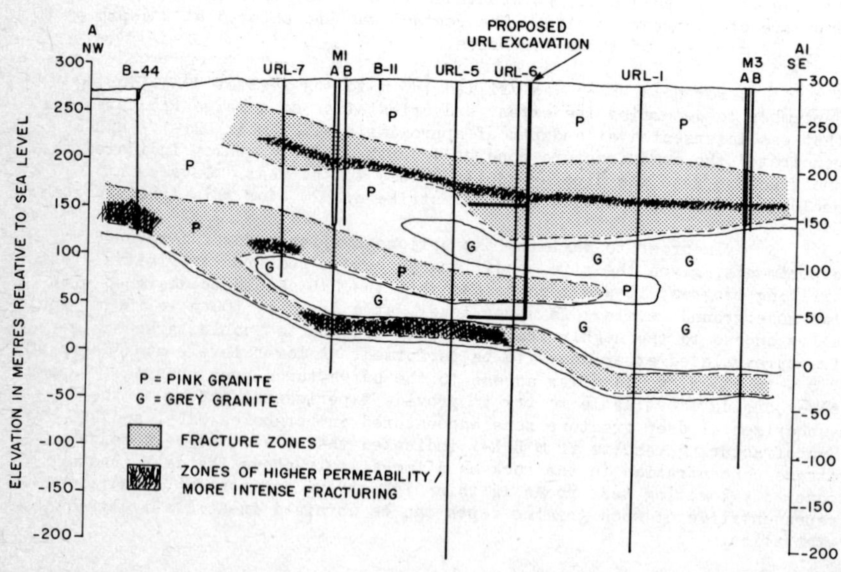

Figure 6: Geologic Cross Section Showing Proposed Location of URL Excavation

During May of 1982, an exploratory shaft pilot borehole (URL-6) was drilled near the location proposed for the URL shaft, to confirm the geologic conditions and to provide the detailed engineering information required to plan the shaft construction. URL-6 was drilled vertically to a depth of 400 m, using continuous NQ wireline coring equipment. Detailed core logging, in-hole surveys, and hydrogeological measurements were performed in this borehole to locate and orient lithologic units and fracture zones. Hydrofracturing tests were also undertaken to establish if the magnitude and orientation of the in situ stress field are the same as those measured in URL-1 by hydrofracturing methods.

After the completion of URL-6, an additional shallow borehole (URL-7) was drilled to investigate the western extent of the fracture zone that was penetrated by URL-1, URL-5 and URL-6. This borehole was drilled to a depth of 200 m in the up-dip direction of the fracture zone in URL-5.

Figure 4 shows the locations of the diamond-drilled boreholes at the URL site. Numerous geophysical surveys were conducted in these boreholes, including television and acoustic television logging, electrical, radioactive, sonic and caliper logs, geothermal profiles, seismic tube wave surveys and full acoustic waveform logging, all of which aided in characterizing the variations in lithology and fracturing that were present in the boreholes. Figure 5 presents an example of typical fracture location data obtained from the borehole core logging and television logging surveys.

A high resolution seismic reflection survey was conducted from the surface over the north-central region of the URL lease area to detect subhorizontal fracture zones. The preliminary interpretation of results has traced the deep fracture zone that was intersected by boreholes URL-1, URL-5 and URL-6.

Figure 6 illustrates a vertical section through the northeast portion of the URL site that presents the lithologic and fracture information obtained from the borehole drilling programs. Some information on this section was obtained from a series of percussion-drilled boreholes, which are being drilled for hydrogeologic monitoring purposes (see later discussion).

Hydrogeological Investigations

The URL site offers researchers a unique opportunity to monitor the perturbation created in the groundwater regime by the construction of the URL shaft and underground workings. The hydrogeology studies that are underway at the URL site have several objectives:

(1) To define the natural undisturbed physical and chemical characteristics of the groundwater regimes within the unconsolidated overburden deposits, the shallow bedrock and the deep bedrock.

(2) To develop and install a hydrogeologic monitoring system in boreholes at the URL site to monitor natural variations in groundwater parameters, in order to establish base-line information prior to the construction of the URL excavations in 1984.

(3) To predict the influence of the URL construction on the surrounding groundwater regimes using computer models that incorporate all the available preconstruction hydrogeologic data.

(4) To monitor the perturbations of the physical and chemical groundwater conditions resulting from the construction of the URL. These data will be compared to earlier predictions and used to validate the computer models that describe the movement of groundwater in large volumes of plutonic rock.

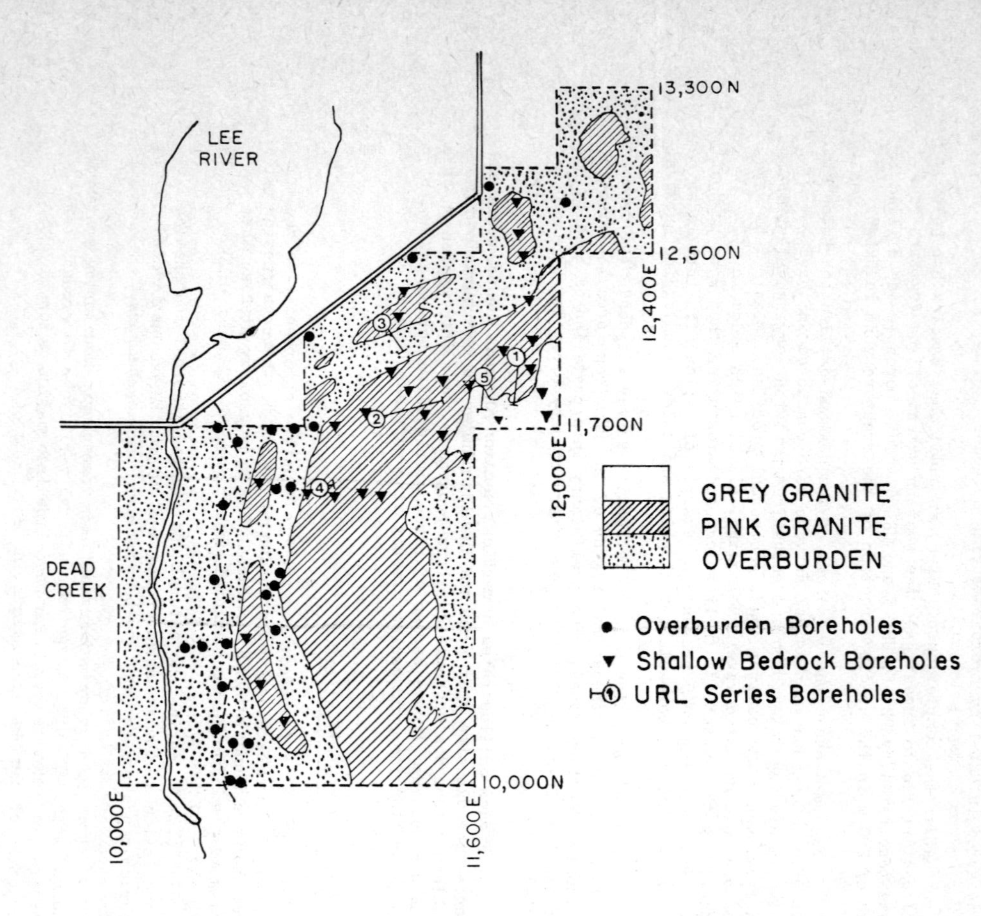

Figure 7: Surficial Geology and Shallow Groundwater Monitoring Network at the URL Site

Shallow Groundwater

At the URL site, the shallow groundwater system has arbitrarily been considered to extend to a depth of 70 m below ground surface. Groundwater conditions in this shallow system need to be studied in detail to define the configuration of the water table, which is the upper boundary in any model of the site, as well as to determine when, how and how much groundwater recharge/discharge occurs.

Most of the depressional areas of the URL site are infilled with glaciolacustrine clays, silts and tills. Numerous shallow test holes (Figure 7) have been drilled into these unconsolidated deposits to determine their thickness and stratigraphy. Generally these materials are 2 to 10 m thick and they consist predominantly of clays and silts. Occasionally, however, a thin layer of glacial till occurs between the clay and the underlying bedrock. Fifty-eight water-table wells and piezometers have been installed in the unconsolidated deposits to measure groundwater conditions. Response tests have been performed in these installations, and the results indicate that the permeability of the unconsolidated deposits can range from 1×10^{-3} $cm.s^{-1}$ for some of the glacial tills to 1×10^{-9} $cm.s^{-1}$ for the clays. Most values are of the order 1×10^{-7} $cm.s^{-1}$. Water levels have been measured weekly in these devices to observe seasonal and other natural fluctuations in the groundwater levels.

A series of 30 boreholes (named the B-series boreholes) was drilled during the winter of 1981 to study the groundwater conditions within the shallow bedrock of the URL site. These boreholes were NQ size (76 mm) cored holes, and were drilled to depths of 10 m to 15 m. Two of the boreholes were extended to 40 and 60 m, respectively, to probe geologic conditions along the western side of the URL site. The cores from the boreholes were logged to locate any fractures intersected by the boreholes. This information was used to select the locations for water-table wells and piezometers, which were installed in these boreholes to make groundwater measurements. Water levels in these installations have been monitored on a regular basis since their completion. The wells and piezometers will be used to define the water table in the bedrock and to determine the nature of groundwater recharge, which occurs as vertical infiltration through the fractures in the bedrock surface at the URL site.

Deep Groundwater

The physical and chemical characteristics of the deep groundwater regime (below a depth of 70 m) at the URL site have been studied from measurements made in the deep exploration diamond-drilled boreholes, drilled during the 1981 and 1982 field seasons.

Hydrogeologic measurements were made in these boreholes by AECL staff during the drilling program to determine the undisturbed hydraulic pressure conditions within the rock mass. After the boreholes were completed and logged by various methods, detailed hydrogeologic tests were performed at selected fracture intervals to define the physical and chemical groundwater characteristics. Figure 8 illustrates typical hydrogeologic information obtained from the borehole investigations.

After detailed hydrogeologic testing in the URL Series boreholes, temporary PIP packers (production-injection packers) were installed in the boreholes to seal off undesireable annular flows. In three of the boreholes, the topmost PIP was connected via a standpipe to surface (refer to Figure 8). This type of installation is effectively a temporary piezometer that allows the zone below the upper PIP to be continuously monitored until the PIPs are removed. The following zones were completed in this manner:

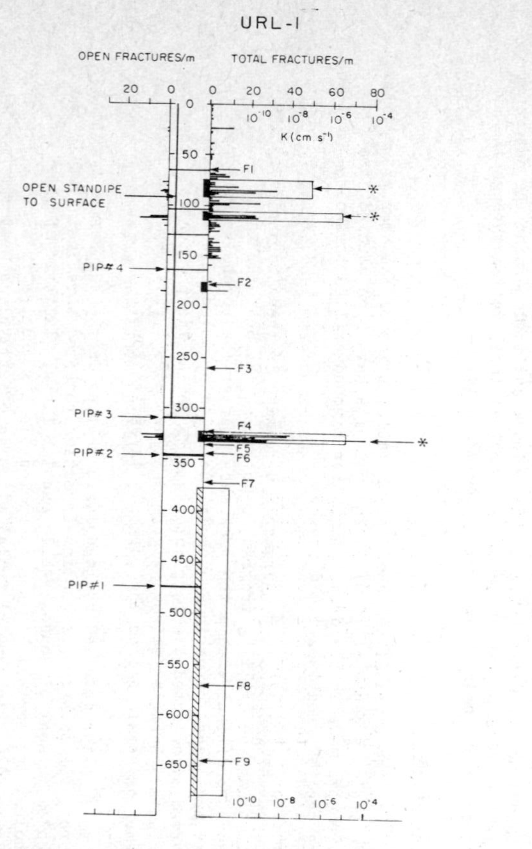

Figure 8: Typical Hydrogeologic Testing Information from URL-1

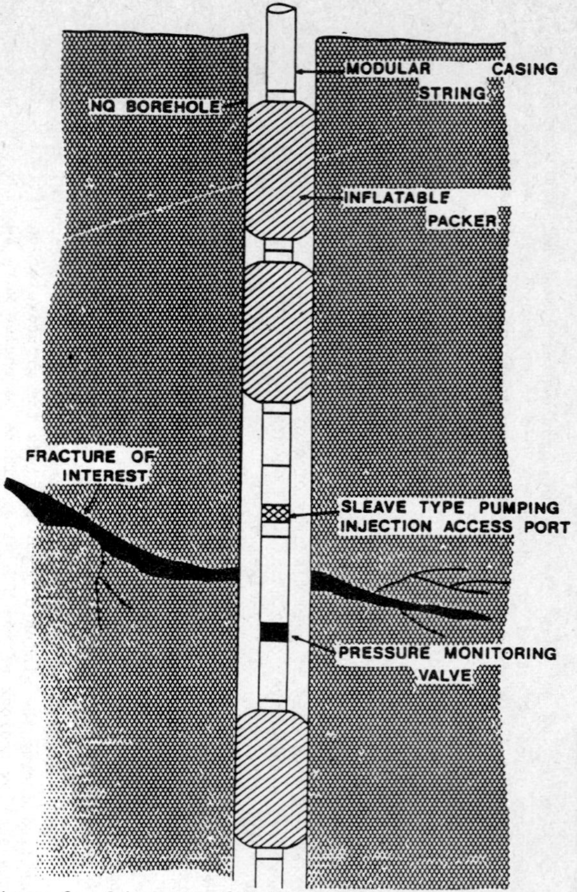

Figure 9: Schematic of Multiple Interval Borehole Completion System for Hydrogeologic Monitoring

URL-1	324.5 m - 331.5 m	deep fracture zone
URL-2	101.0 m - 180.0 m	
URL-5	250.2 m - 269.4 m	deep fracture zone.

Water samples were collected from selected intervals in the boreholes to determine the chemical composition of the groundwaters. Major dissolved ions and gases have been analyzed in these samples. Preliminary hydrogeochemical information from these samples indicates that several different groundwater flow regimes exist at the URL site. Shallow groundwaters are generally low in total dissolved solids and of the Ca-HCO_3 type. Occasionally, within shallow, highly fractured and altered zones, Na-HCO_3 type water is present. Water within the deeper fracture zone is a Ca, Na-Cl type, and chloride contents up to 5000 mg/L have been found in some boreholes. Because of the chemical character of the chloride-enriched groundwater within the deeper fracture zone, it is an excellent tracer. Similar water has been sampled from shallow depths in boreholes in the northern portion of the URL lease areas. This indicates that some of this groundwater is moving up the fracture zone, from a Na, Ca-Cl saline source at depth to shallow depths. The water at depth within unfractured, low-permeability portions of the URL lease area is saline, but this water is of the calcium chloride type and is distinctly different from that associated with high-permeability zones.

Hydrogeologic Monitoring

Hydrogeologic monitoring equipment has been developed for installation in the network of NQ size (76 mm) boreholes at the URL site. The equipment consists of a stainless steel multiple-interval casing system, as illustrated in Figure 9.

During 1982, a series of large diameter (156 mm diameter) percussion-drilled boreholes are being drilled at the URL site to complete the hydrogeologic monitoring network. These are the M-Series boreholes shown on Figure 10. Many of these boreholes are being completed, as illustrated in Figure 11, to ensure that bottom-hole single-packer piezometer completions can be installed if multiple-interval casing systems are not used in them.

Hydrogeologic and geologic conditions are being determined in the percussion-drilled boreholes as they are completed, and this information has been used to help prepare the vertical section of the north portion of the URL lease area, as shown in Figure 6. These boreholes have been located to provide additional information regarding the three-dimensional orientation of the deep fracture zone that was previously intersected in the north portion of the URL site by URL-1, URL-5 and URL-6. Preliminary results indicate that the fracture zone is a continuous region, approximately 10-15 m thick, of intense fracturing in the northeast portion of the URL site. Permeability characteristics within the zone itself are highly variable and values ranging over several orders of magnitude have been determined. A region of extremely high permeability exists in the fracture zone in the north and east, whereas low permeabilities are present in the south and east (see Figure 12). The fracture zone dips up to surface in the western portion of the URL lease area, and it appears to be hydraulically well-connected to shallow portions of the rock mass in the north and west sections of the URL lease area. Groundwater of the Na, Ca-Cl type appears to be migrating up the fracture zone from a deep-seated saline source in the northeast portion of the URL site. This water mixes continuously in the fracture zone with dilute Ca-HCO_3 type water, which is vertically recharged locally in the south from the surrounding rock mass to form a more dilute Ca, Na-Cl water. Leakage out of the fracture zone occurs upward into the shallow groundwater flow systems in the north portion of the URL lease area.

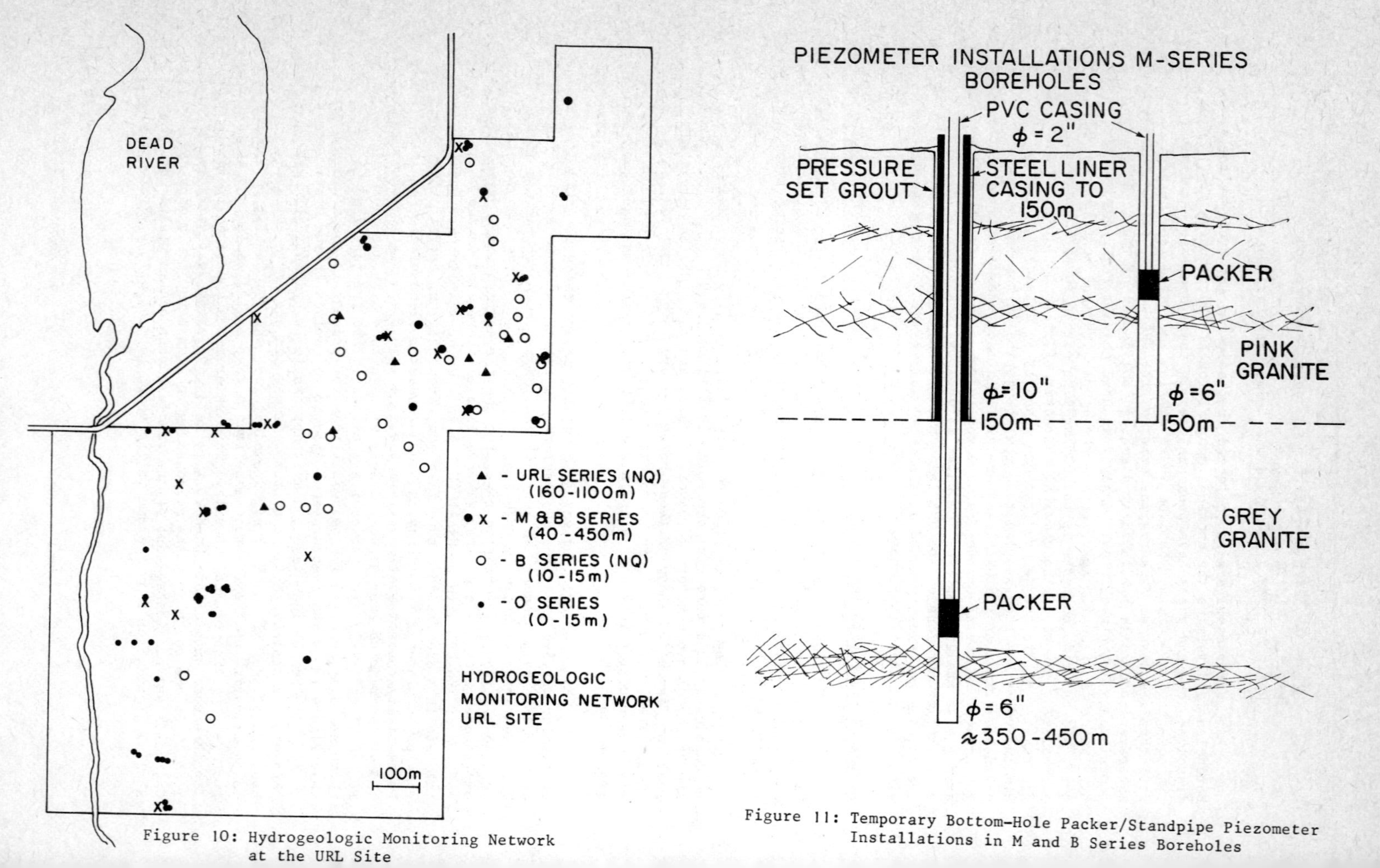

Figure 10: Hydrogeologic Monitoring Network at the URL Site

Figure 11: Temporary Bottom-Hole Packer/Standpipe Piezometer Installations in M and B Series Boreholes

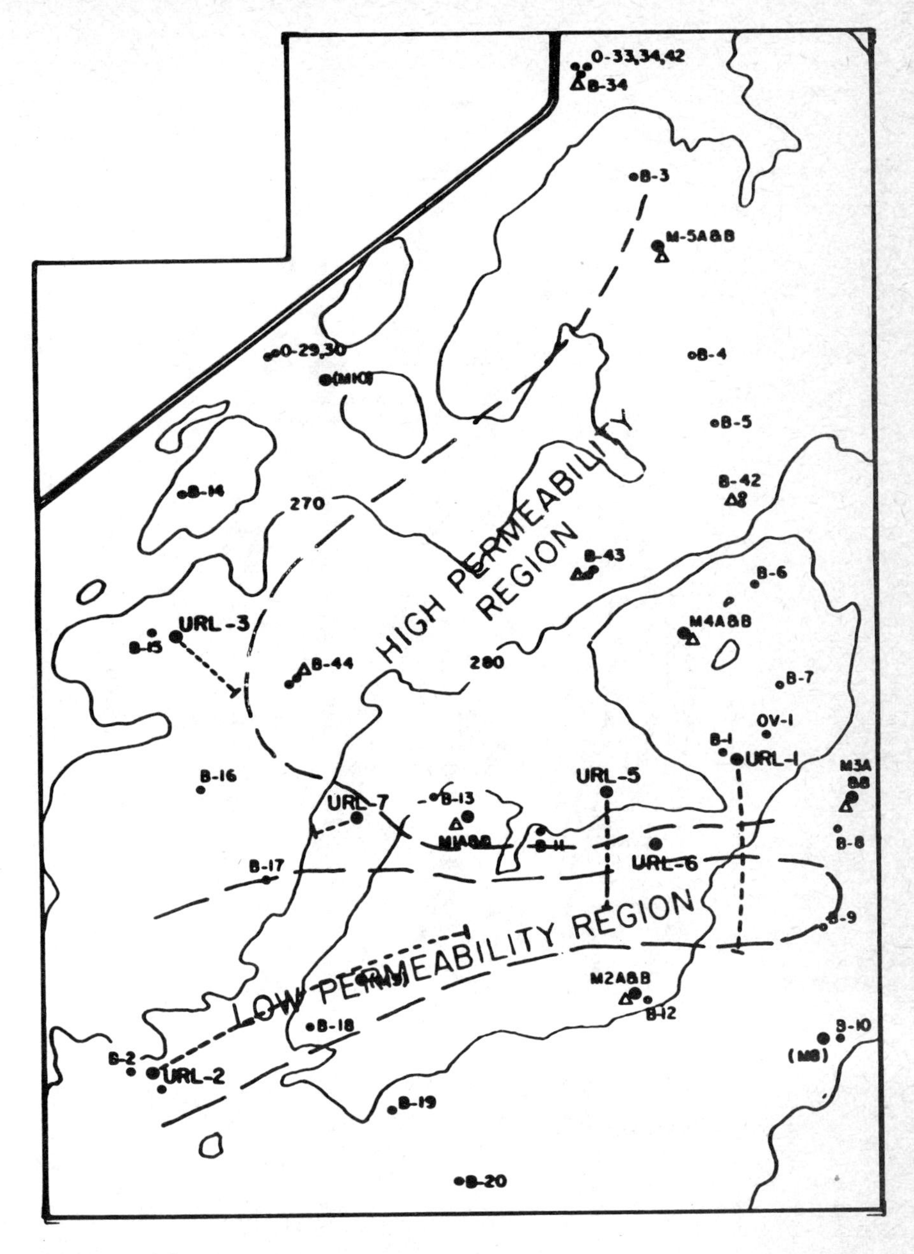

Figure 12: Preliminary Interpretation of Permeability Distribution in Deep Fracture Zone - NE Portion of URL Site

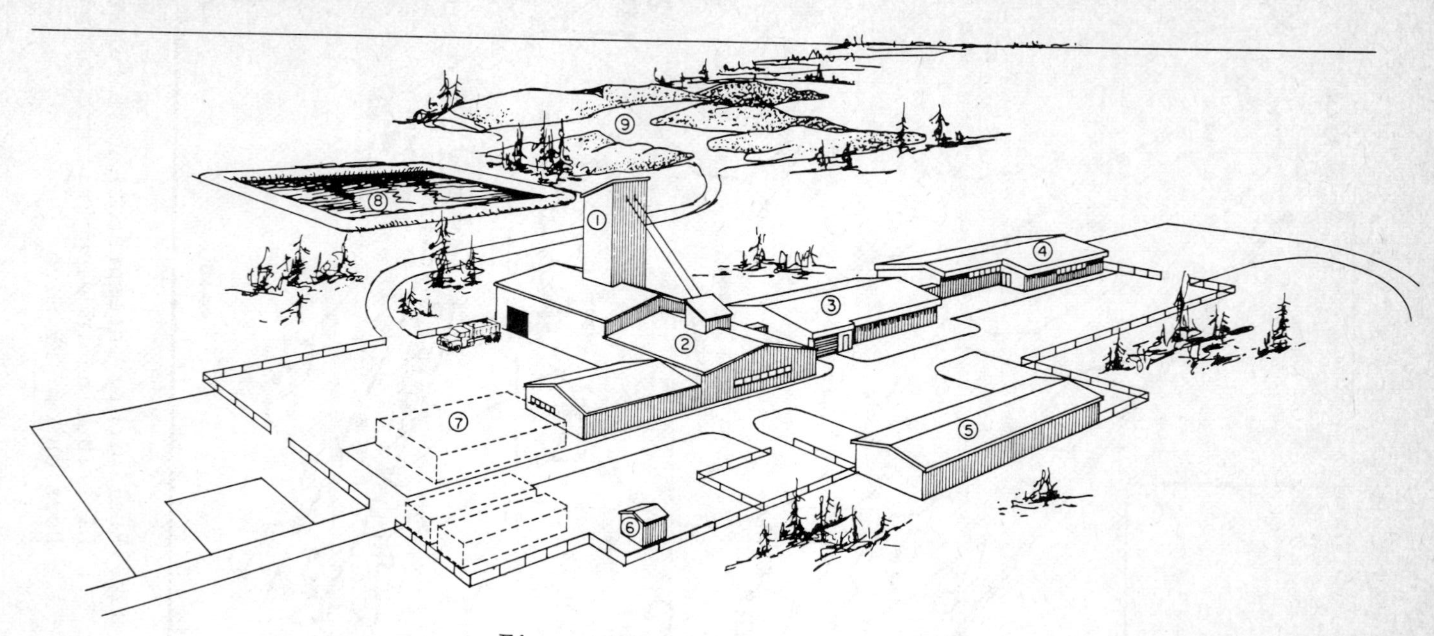

Figure 13: URL Surface Facilities

URL DESIGN AND CONSTRUCTION

When the URL site evaluation program had provided the initial general interpretation of the geologic environment of the URL lease, a site for the URL was selected and the design process begun. The criteria applied to the design are that (1) access be provided to representative geologic environments for experiments and (2) the underground arrangement suits the proposed experiments. The designs for the surface and underground facilities have evolved through extensive discussion amongst the experimenters, the URL operations group and the URL design consultants. In the first phase of the design, the design consultants reviewed all identified needs and produced a preliminary design and construction cost estimate. This proposal was subjected to extensive review by other design consultants, experimental program leaders, and invited outside groups. The preliminary design proposal was revised, based on this review, and detailed design has begun [3].

The detailed design will provide packages of tender and contract documents covering the URL facilities shown in Figures 13 and 14. The URL surface facilities will provide the services necessary to maintain the underground operation and facilitate the development and execution of the experiments. Major installations will include an Office and Public Affairs Building, a Laboratory and Maintenance Building, a Headframe and Hoist Complex and a Vehicle Garage and Drill Core Storage Building. In addition to the permanent facilities, temporary buildings will be installed to house the contractor's workshops, offices and washrooms.

The URL underground facilities comprise a 250-m access shaft, a ventilation raise, and experimental test areas on two levels. The shaft will be rectangular, 2.8 m by 4.9 m, and timber lined. The upper test level, at a depth of about 130 m in somewhat fractured rock, will consist of a 100-m access drift and two main test rooms. The main test level, at a depth of about 240 m in very competent rock, will consist of a 145-m access drift, a shop and instrumentation test room, three parallel rooms for multicomponent tests, a hydrogeology and geochemistry test area, and an area designated for excavation response and shaft-sealing experiments. Provision will be made for future excavations as required, including access to a fracture zone located below the lower level, where experiments in a very fractured rock mass may be undertaken during the operating phase.

The URL construction schedule, shown in Figure 15, has been developed to minimize disturbance to the hydrogeologic system prior to 1984. As well, sufficient time will be allowed to develop and implement the tests to be conducted during construction. Underground development of the shaft and of the levels and the raise bore will be separated by a period of about eight months, during which the shaft and shaft stations will be characterized. No construction activities are scheduled during this period. The major schedule milestones are:

Begin site preparation	1982 October
Construct surface buildings and collar the shaft	1983 April to November
Excavate the shaft	1984 January to August
Geotechnical characterization of the shaft	1984 September to 1985 April
Level development and raise boring	1985 May to 1985 December

URL UNDERGROUND EXPERIMENTAL PROGRAM

The underground experimental program subdivides chronologically into construction, geotechnical characterization and operating phases. The

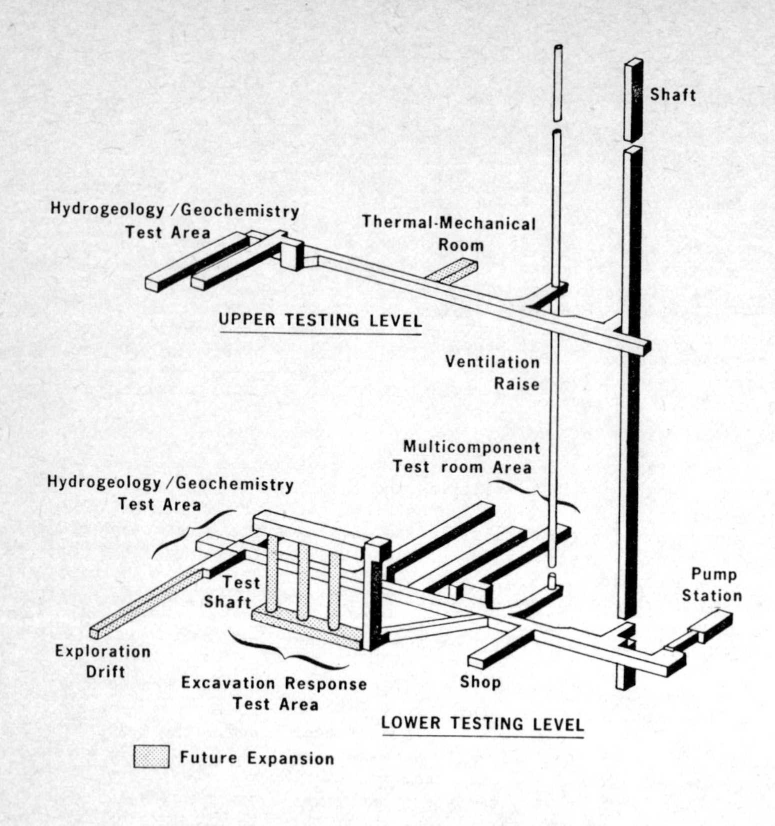

Figure 14: **URL UNDERGROUND LAYOUT**

Figure 15: **URL DESIGN & CONSTRUCTION SCHEDULE**

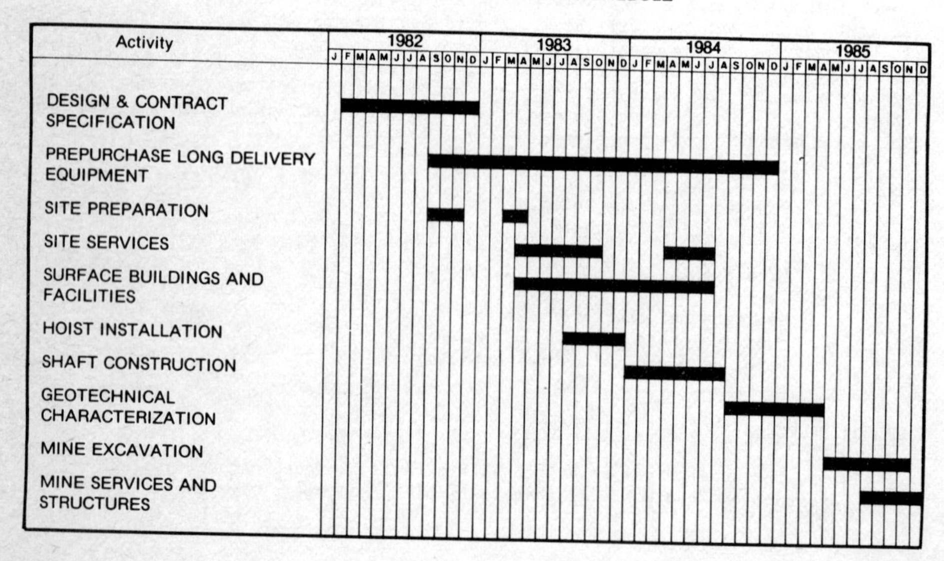

construction-phase activities are those in which excavation is an integral component, or for which adequate access can only be achieved during construction. The geotechnical characterization on the test levels is done after construction; it completes construction-phase experiments and characterizes areas where operating-phase experiments will be undertaken. The operating-phase experiments will study complex, coupled conditions that may be encountered in a disposal vault, or will develop and assess techniques that are essential to the feasibility of the disposal concept.

Activities During Shaft Excavation

Detailed geologic mapping of exposed rock in the walls and benches of the shaft will be done daily during shaft sinking. The benches that are exposed and will be excavated prior to the next mapping period may be photographed to provide a permanent record. These data will provide the basis for assessing the geologic predictive capability and the subsurface model developed from the site evaluation program data.

During shaft excavation, geophysical surveys will provide additional information on the structure of the geologic environment. Vertical seismic profiling, using small explosive charges at the base of the shaft monitored by an array of surface geophones, will identify structural features in the rock surrounding the excavated shaft. Radar sounding, VLF-EM, gamma-ray spectroscopy, magnetic susceptibility, pulse EM and gravity techniques may be used to survey for structural features in the rock below the shaft in the path of the excavation. The value of these techniques will be assessed and any additional structural information will be used in assessing and improving the subsurface geologic model.

During shaft sinking, the rock-mass response to excavation will be measured at four depths. For each measurement, two multi-point extensometers will be installed horizontally in two directions immediately above the bottom of the shaft. Each extensometer will have several reference anchors and will be installed in 15-m-long NX or EX boreholes. The displacement of the anchors will be monitored as each subsequent shaft bench is excavated. These data will allow calculation of the effective rock-mass modulus at various locations.

At each of the four locations where the rock-mass response to excavation is measured, the near-field hydrogeologic system response to shaft excavation will be monitored. A single horizontal NX hole, 15 m long, will be drilled into the rock mass immediately above the shaft bottom. The hydraulic head at various locations in the borehole will be monitored as shaft sinking proceeds. This will provide data on the rock-mass dewatering near the excavation.

The engineered blasting program during shaft excavation involves (1) detailed review and excavation damage estimates for the controlled (i.e,. normal construction) blast round, (2) design of careful (i.e., experimental) blast round to minimize wall damage in two 20-m shaft sections, along with estimates of damage to be expected, and (3) field participation at the time the experimental zones are excavated. Two experimental zones are planned: one just below the upper working level and one just above the lower working level in the shaft. These zones should provide excellent rock wall quality and may be the locations for shaft-sealing tests during URL decommissioning. The actual excavation damage from controlled and careful blast rounds will be measured and compared with the estimates.

Water samples will be taken at any location in the URL shaft where groundwater flows from fractures. A sampling device is being developed for installation on flowing fractures above the excavation zone to collect uncontaminated water samples; these will be analyzed to determine their composition, age and origin. These data will add to the hydrogeologic monitoring data base and aid in understanding the hydrogeologic flow systems in the URL area of the Lac du Bonnet batholith.

Activities During Geotechnical Characterization of the Shaft

The excavation damage will be assessed during the geotechnical characterization phase, which will extend work by the Colorado School of Mines [4]. Horizontal holes will be drilled into the shaft walls, possibly two 15-m-long parallel holes into each wall. Four techniques are planned for assessing the extent of damage in the rock mass. The core obtained from these holes will be fracture logged. The holes will be tested, individually or cross hole, with permeability, modulus and ultrasonic equipment. Testing at regular intervals, say every 0.25 m, will indicate the variation in the rock-mass property variation with distance from the shaft wall and, thus, the extent of excavation damage.

At four locations in the shaft and at the shaft station for each working level, the in situ stress in the rock mass will be measured using United States Bureau of Mines (USBM) and Council of Scientific and Industrial Research, South Africa, (CSIR) strain cells. At each measurement location, a series of five USBM and two or three CSIR overcore stress measurements will be taken horizontally in two directions over a distance of 15 m. These data will provide a basis for comparison with hydraulic fracturing stress measurements done in boreholes from the surface. The orientation of the stress field will be taken into account in selecting the final orientation of the access tunnel for each URL test level.

The far-field hydrogeological field program being conducted in boreholes drilled from the surface will be augmented by three 200-m-long NX boreholes drilled subhorizontally from each shaft station at 120° spacing. This will provide a horizontal monitoring array at depths of 130 m and 240 m for assessing the vertical component of the hydrogeological flow systems. This network will also provide geochemical data and information on the variation in hydraulic gradients towards the shaft.

In order to confirm the suitability of the rock mass at the depth and orientation selected for the main access tunnel on each working level, pilot boreholes will be drilled. The core will be logged in detail to ensure there are no unexpected geological or hydrogeological conditions in the areas to be developed for test rooms. The core fracture information will be extrapolated to the scale of the access tunnels and test rooms. The accuracy of this extrapolation will be assessed during geologic mapping of the horizontal excavations.

Activities During Horizontal Development

During the excavation of the shop/instrumentation test area on the lower level of the URL, a hydrogeologic assessment will be done. Prior to excavation, three horizontal NX boreholes will be drilled at the room center-line height at three distances outside the boundary of the room excavation (say 1.5, 3.5 and 6 m). Extensive permeability, or hydraulic conductivity, testing in single holes and cross hole will provide a data base on the properties of the rock prior to room excavation. During the excavation of the shop, the changes in hydrogeologic parameters will be monitored continuously to assess the effect of excavation on the near-field.

During the excavation of the testing levels, the walls of tunnels and test rooms will be geologically mapped. The excavation sequence is being arranged normally to provide undisturbed access to one excavation area every day. These data will be used to assess and improve the geologic model of the URL site. As well, the face of selected excavation headings will be mapped daily, and the geology of subsequent faces will be predicted.

During level development, the geophysical techniques found to be effective during shaft excavation will be used to expand the data base on the geologic environment in advance of the excavation. The effectiveness of the techniques will be compared with that of radar surveys.

The response of the rock mass around the tunnels to excavation will measured at two locations on each level. The technique is similar to that to be used during shaft excavation. These tests will provide an estimate of the effective rock-mass modulus for the rock mass surrounding each test level.

A major excavation response and engineered blasting test, similar in some aspects to the mine-by experiment conducted by Lawrence Livermore National Laboratories in the Climax facility at the Nevada test site [5] are planned for the test shaft area in the URL. In this experiment, the access tunnel and upper test-shaft access room will be excavated before the lower test-shaft access room. The volume of rock through which the lower access room will be excavated will be extensively characterized and instrumented in three zones. The undisturbed hydrogeologic properties will be established and extensometers will be installed around the path of the room in each zone. As the lower access room is excavated, the response of the rock mass will be measured and compared with mathematical predictions, calculated using effective rock-mass properties determined in shaft and tunnel rock-mass response tests. The lower access tunnel will be excavated using three very different blast-round designs, and the extent of excavation damage in each zone will be assessed.

The engineered blasting program during horizontal development involves (1) detailed review and excavation damage assessment for the controlled, or construction, blast rounds, (2) design of three distinctively different blast rounds for excavating the lower test-shaft access room, along with estimates of the extent of blast damage for each design, and (3) field participation at the time the experimental zones are excavated. The three blast-round designs will each be used to excavate a third of the lower test-shaft access room, and will be designed to provide a different degree of excavation damage. The actual extent of blast damage will be assessed during the geotechnical characterization of the levels, following completion of URL construction. Another aspect of the engineered blasting program is the application of performance criteria to the excavation done by the URL construction contractor. Performance criteria, requiring a reasonable degree of control, will be applied to the standard excavation blasting, and much more stringent criteria will be applied to the excavation of one multicomponent test room on the lower level. The extent of excavation damage under both criteria will also be assessed.

Groundwater samples for chemical analysis will be taken from any locations where seepage or flow of water occurs. Sampling devices are now being developed for this purpose. The data from these analyses will be added to the data base on the groundwater chemistry at the URL lease area.

Activities during Geotechnical Characterization of Test Levels

Following completion of the URL primary construction contract activities, when AECL has assumed operational control of the URL, a series of characterization tests will be conducted on the upper and lower testing levels. The general objective of these tests will be to complete the assessment of the construction-phase experiments and to characterize areas in which future experiments will be undertaken. The major activities during this phase are discussed in the following sections.

Excavation Damage Assessment

When excavation of the URL is complete, an assessment of the extent of damage caused by the various excavation techniques (blasting and raise boring) will be undertaken. At selected locations in the ventilation shaft, access tunnels and test rooms, the variation in fracture density, permeability, and acoustic and mechanical properties with distance into the rock mass from the excavation wall will be assessed. The data will be used in assessing the relative effectiveness of the raise-boring technique and various blast-round designs for controlling wall damage, and will provide input to the vault-sealing program (grouting requirements and design parameters for shaft and drift seals). The data will also be used to assess the effectiveness of the engineered blasting methodology at predicting excavation damage.

Stress Field Mapping

The stress field in the rock mass surrounding both URL levels will be mapped by a series of two- and three-dimensional stress measurements, taken at selected locations in the horizontal excavations. Both USBM and CSIR strain cells will be used in overcore stress determinations in boreholes drilled horizontally and vertically at several locations.

Thermal and Mechanical Properties of the Rock Mass

In addition to a knowledge of the ambient stress field, the experimenters and analysts must understand the in situ mechanical and thermal rock-mass properties. If these properties vary with stress and/or temperature, the variations must be quantified. In the URL, a block test and a heater test are planned to provide this information.

A block test will be used to provide mechanical, thermal, and coupled thermal-mechanical response data for a large block of rock containing typical fracturing. Information on hydraulic conductivity and mass transport along a discrete fracture may also be obtained, if the rock being tested contains a suitable fracture.

Block tests have been conducted at the Colorado School of Mines [6] and are being run at the Near Surface Test Facility on the Hanford Reservation, Washington. A variation of this test has been proposed for the URL by Golder Associates [7]. In this proposal, a single slot, at least 6 m long, is drilled vertically down from the upper-level test drift to a level below the floor of the instrumentation drift. Four 2-m-square, high-capacity (15 MPa) flatjacks are grouted into the slot at the elevation of the sublevel drift. Twelve heaters and four multipoint extensometers are installed horizontally from the sublevel drift into the volume of rock under test. Vertical monitoring holes for stress and permeability measurements are drilled vertically into the test rock. Extensive scoping calculations are still necessary to ensure that the modified block concept will provide the desired data.

A single-heater test is planned as a means of determining the in situ thermal properties of the rock mass and of assessing the accuracy of the thermal-mechanical models. Since similar experiments have been done in the Stripa Mine in Sweden [8] and in the Near Surface Test Facility (NSTF) at Hanford [9] with varying degrees of success, these experiments will be used to confirm that the response of the Lac du Bonnet batholith can be predicted and to confirm the values of thermal properties necessary to model later experiments.

For this test, the heater will be installed in a borehole in the floor of a test room and surrounded by instruments to monitor temperature distributions, displacements, and stress-field variations that occur within the rock during the tests. In addition, monitoring holes will be provided for geophysical and fluid conductivity measurements, to be made at various intervals during the test.

Operating-Phase Experiments

The experiments that will be designed and operated during the life of the URL will involve complex, coupled conditions that may be encountered in a disposal vault, or will develop and assess techniques that are essential to the feasibility of the disposal concept. These experiments will not be designed in detail until parameters such as waste package geometry and heat output, buffer and backfill material composition and thickness, and the in situ conditions and properties of the rock mass are known. In the following sections some concepts for operating-phase activities are briefly discussed.

Heater/Buffer Experiments

A heater/buffer experiment will be initiated in the URL similar to the buffer/mass test presently underway in the Stripa Mine [10]. It will provide data for assessing the mathematical models of the heater/buffer/rock-mass system and the buffer/rock-mass/groundwater interactions.

Hydrogeology and Geochemistry Experiments

Several hydrogeology and geochemistry experiments are being proposed in the URL program to determine the directional hydraulic conductivity and mass-transport properties of the rock mass under a variety of conditions. Hydrogeologic and geochemical test areas will be provided on both levels in the URL as stations from which test boreholes can be drilled. Water or chemical tracers will be injected in some boreholes, while monitoring the conditions in selected test intervals of surrounding boreholes.

The hydraulic conductivity experiments may be run at ambient temperature, and then in an electrically heated rock mass, to assess the effect of a thermal field. The power level of the heaters will be regulated to vary the rock-mass temperature and the thermal gradient, to provide data for a range of conditions. The tracer experiments will follow, with test conditions ranging from ambient to maximum temperature.

Retardation experiments will be run in well characterized fractures in the rock mass surrounding the URL. These experiments, using chemical tracers and possibly radioactive tracers, will provide a means of assessing the accuracy of mathematical models that simulate mass-transport phenomena.

Vault Sealing

The vault-sealing program includes the development and testing of buffer and backfill materials, and shaft and drift seals. The URL will provide a representative geologic environment to evaluate various concepts. Within the URL, there will be areas for emplacing and testing buffer and backfill materials and drift and shaft seals.

Moisture Balance or Macro-Permeability Experiment

The hydraulic conductivity of a large volume of rock surrounding a portion of the URL will be measured. The various hydrogeologic and geochemical monitoring holes in the URL will provide information on the hydraulic gradient in the surrounding rock. The amount of moisture inflow to an experimental drift, or possibly to the upper or lower testing level, will be measured by monitoring the liquid inflow and outflow and the moisture content in the ventilation air flow. The moisture inflow rate and the hydraulic gradient existing around the test excavation will provide an estimate of the bulk rock permeability. These data will be used in the development of the geosphere model in the environmental and safety assessment studies.

Large-Scale Multicomponent Experiment

As the program progresses and the less complicated experiments are essentially completed, one or more large-scale multicomponent experiments will be designed. Prior to such experiments, the waste container geometry and thermal output, and the buffer and backfill materials and thicknesses to be used in the Canadian disposal program, must be specified. These experiments could simulate, at full scale, a section of a disposal vault room, with heaters designed to simulate, geometrically and thermally, the proposed waste containers.

Tests could be run to determine the thermal and mechanical responses of the test room to the heaters and buffer material installed in boreholes in the floor. The room could then be backfilled with the proposed backfilling material, with instruments installed to monitor the temperature response and the rewetting transient in the room.

Such an experiment would provide a thorough test of the modelling and assessment capability to be developed during the next 10 years.

URL PROGRAM DEVELOPMENT

This paper has discussed the URL experimental program. Detailed planning of many experiments is now underway. The first task is to document detailed plans for construction-phase experiments. This document is scheduled for completion by early 1983.

Subsequently, the detailed plans for geotechnical characterization of the levels and an overview of the operating-phase experiments will be compiled. The detailed planning of the various phases of the URL program will remain flexible as long as practical, to allow incorporation of experience gained in underground testing at other sites.

REFERENCES

1. Simmons, G.R. and Soonawala, N.M. (editors), 1982, "The Underground Research Laboratory Experimental Program", Atomic Energy of Canada Limited Technical Record*, TR-153.

2. Soonawala, N.M., Davison, C.C., and Brown, A. "Geological, Geophysical and Hydrogeological Investigations at the Site of the Planned Underground Research Laboratory", IN Proceedings of the Canadian Nuclear Society International Conference on Radioactive Waste Management, Winnipeg, 1982 September, in preparation.

3. Peters, D.A. and Simmons, G.R., "The Underground Research Laboratory Design" IN Proceedings of the 14th Information Meeting of the Nuclear Fuel Waste Management Program (1982 General Meeting); Atomic Energy of Canada Limited Technical Record*, TR-207, in preparation.

4. Hustrulid, W.A., Cudnick, R., Trent, R. and Holmberg, R., 1980, Mining Technology Development for Hardrock Excavation, IN Proceedings of the Workshop on Thermomechanical-Hydrochemical Modeling for a Hard Rock Waste Repository, Lawrence Berkeley Laboratories Report LBL-11204, pp. 56-61.

5. Heuze, F.E., 1981, "Geomechanics of the Climax "Mine-by", Nevada Test Site", IN Proceedings of the 22nd U.S. Symposium on Rock Mechanics, pp. 428-434, Massachusetts Institute of Technology, 1981 June 28-July 02.

6. Voegele, M., et al, 1981, "Site Characterization of Joint Permeability Using the Heated Block Test", IN Proceedings of the 22nd U.S. Symposium on Rock Mechanics, pp. 120-127, Massachusetts Institute of Technology, 1981 June 28-July 2.

7. Golder, H.Q., and Associates Limited, 1980. "A Conceptual Design Study for an Underground Research Laboratory", H.Q. Golder and Associates Limited Report No. 791-1187, 1980 December (also issued as Atomic Energy of Canada Limited Technical Record*, TR-160, 1982).

8. Schrauf, T., Pratt, H., Simonson, E. et al, 1979, "Instrumentation Evaluation, Calibration, and Installation for Heater Experiments at Stripa", Lawrence Berkeley Laboratories Report LBL-8313.

9. Basalt Waste Isolation Program, 1980, "Near Surface Test Facility Program (Phase 1 and Phase 2)", Report No. BWIO2T P0101 Rev. 1.

10. Carlsson, H., 1980, "The Stripa Project, A Multilateral Project in the Management of Radioactive Waste Storage", Subsurface Space, Rock Store 80, Vol. 2, Stockholm, Sweden, Pergamon Press, pp. 819-828.

*Unrestricted, unpublished report available from SDDO, Atomic Energy of Canada Limited Research Company, Chalk River, Ontario KOJ 1JO.

PROPOSALS FOR IN-SITU RESEARCH IN THE PROPOSED LABORATORY
AT GRIMSEL IN SWITZERLAND

E. Pfister
Nagra, National Cooperative for the Storage
of Radioactive Waste
Baden, Switzerland

ABSTRACT

The National Cooperative for the Storage of Radioactive Waste, Nagra, has proposed an underground research laboratory in the Grimsel area of the Swiss Alps. Some site characteristic data are presented. A series of in-situ experiments is being developed in order to the needs of the Swiss waste storage concept. Several of the proposed experiments are discussed in this paper.

INTRODUCTION

In Switzerland responsibility for radioactive waste disposal lies directly with the producers of wastes. Accordingly, utilities with nuclear power interests together with the Federal Office of Public Health (which is responsible for industrial, medicinal and research wastes) have formed Nagra (National Cooperative for the Storage of Radioactive Waste) to perform the necessary project work.

The work at Nagra must include projects for disposal facilities for all types of active wastes. Most effort is devoted to geological, geophysical and hydrological field work.

The disposal concept in Switzerland is directly affected by geography and geology which determines the host rocks available. For HLW attention is centered upon the granitic crystalline basement underlaying the country. The fact that this rock is reachable only in a limited area which is overburdened by hundreds of meters of sedimentary rocks has the direct consequence that the region for site investigation is fixed. Besides an extensive geophysics program, a series of 12 deep drillings, each penetrating 1000 m into the crystalline basement, began on the northern part of Switzerland in October 1982.

The mechanical, thermal, hydrogeologic and geochemical properties and the mechanisms that control the behaviour of deep cristalline formations must be understood to develop mathematical models. To meet the needs of the Swiss program, an underground research laboratory is being planned in the Alps, near Grimsel, about 50 km south of Lucerne.

OBJECTIVES OF THE ROCK LABORATORY

Testing in an underground environment is required for the development and assessment of techniques for site selection, design, construction and environmental assessment. There is a practical limit to how much can be learned by borehole investigations, laboratory experiments or computer simulations, all of which are now underway in the broad waste-disposal program of Nagra. Although Nagra is a partner in the Stripa project in Sweden it was felt that more direct experience in underground experimentation is required. The Grimsel Rock Laboratory will contribute data and experience in the areas of geophysics, geohydrology, migration of radionuclides, rock mechanics, sealing and thermally induced processes.

The objectives are to test the applicability of results obtained elsewhere, to study aspects specific to Nagra repository concepts and to build up appropriate expertise and practical experience, with a view to later experiments at an actual repository site. The objective is not to characterize the Grimsel site itself since alpine granites are not envisaged as HLW host rocks.

Grimsel can be developed quickly and at a more reasonable cost because a large underground pumped-storage scheme already exists, so that horizontal access to a suitable rock mass with about 450 m overburden is possible.

SITE AND ROCK CHARACTERISTICS

In 1980, 6 horizontal, 100 m long boreholes (Ø = 86 mm) have been drilled from the existing tunnel system. A double tube core barrel was used to achieve a careful drilling. The oriented cores from the drilling were logged using extensive logging procedures. Samples were taken for physical determinations and chemical analysis of the fissure fillings. The boreholes were sealed by a single packer and the development of the hydraulic head was obsrved.

Some of the findings (1), (2) are summarized below.

Petrography

Two principal rock types occur in the test area: Central Aaregranit (with varying amounts of biotite) and Grimsel-Granodiorite both of which are locally traversed by lamprophyre and aplite. They have a different content of plagioclase and alkalic felspar. Generally the granodiorite is the more basic rock. The content of SiO_2 of the granite and granodiorite is 68 - 74 %, 65 - 69 % respectively. There is a gradual change between granite and granodiorite, a clearly defined transition could not be observed. The content of quarz lies between 25 and 50 %.

Beside the more compact rock there are also a few weaker zones such as lamprophyre and aplite veins, greatly tectonized rock (e.g. augen gneiss, mylonite, slawn), and hydrothermally altered rock partly filled with fracture filling minerals.

Jointing

Studies have shown that there are 4 sets of jointing and 3 sets of foliation. One may distinguish between joint systems which already existed before the orogenesis and systems which were developed during the orogenetic movement. Most of the joints are steep, i.e. the dip normally lies around 75° with a lowest value of 50° and a maximum value of 84°.

The degree of jointing is very small. 478 of 600 meters of core did not show any open joints, i.e. about 80 % of the drilled rock is jointless. Only 22 of 600 drill meters (= 3.6 %) showed 5 or more open joints per meter.

It is difficult to judge the length of the joints. In some cases it was possible to corrolate joints over 50 and more meters.

Some of the joints are filled with different minerals like epidote, quartz, dolomite and calcite. It is assumed that the filling material of the joints in general has a higher degree of porosity than the bulk rock.

Hydrogeology

The objective of the hydrogeological investigations was to get an idea of the hydraulic conductivity and piezometric pressure of the rock in order to determine the ground-water flow. The drainage of the boreholes was very low. With one exception of 3.5 l/min, the water outflow was less than 0.1 l/min per 100 m of borehole. The determination of the hydraulic conductivity was done by using injection tests under several different pressures. Only specially selected 5 m sections with a high degree of jointing were tested. The maximum water uptake was 0.75 l/min at 30 bar. The calculated k-value was $10^{-9} \div 10^{-12}$ m/sec. The piezometric head was determined by sealing the holes with a single packer near the borehole opening. The observed pressure heads varied from 12 to 43 bar. In none of the boreholes, which were more or less parallel and about 100 m apart was the same reading obtained.

To document the chemical composition and the age of the groundwater hydrochemical investigations were carried out. Analyses with respect to the following ions were performed: Na, K, Ca, Mg, Sr, Cl, SO_4, HCO_3, PO_4, NO_3, Mn, Fe, Al, SiO_2, F. In addition, HCO_3, O_2, CO_2, T, Eh, pH and the electrical conductivity were determined at the site at each borehole and at different locations in the access tunnel and at the surface. The identification of active flows in the rock mass was done by means of the analysis of tritium, which has a half life of 12.35 years. The observed low content of this isotope (generally $<$1TU) signifies that the water is older than 28 years.

The above mentioned observations lead to the conclusions that the circulation of the water generally occurs slowly on independent flow paths and that there is no continous water table.

Rock mechanics

Rock property data were determined on a total of 30 m of core. In order to check the influence of the anisotropy several small cores were cut out in different directions and tested. The mean values of the unconfined compressive strength were between 120 and 220 MN/m^2. Young's modulus ranges from 42 to 60 GN/m^2. Poisson's ratio varies between 0.25 and 0.27. The shear wave velocity ranges from 2.1 to 3.9 km/sec. The observed values are very typical for crystalline formations.

EXPERIMENTAL PROGRAM

A conceptual experimental program has been mapped out with emphasis on the areas of geophysics, neotectonics, hydrogeology, migration and rock mechanics. The development of the program has been a combined undertaking involving various Swiss organizations, which will work under contract to Nagra, and German groups which will carry out a number of experiments. Organizationally this international cooperation will be on the basis of letters exchanged at government level followed by direct contracts between Nagra, GSF (Gesellschaft für Strahlen- und Umweltforschung) and BGR (Bundesanastalt für Geowissenschaften und Rohstoffe).

In the following sections, the layout and several of the experiments are described.

Layout at the laboratory

The conceptual layout which meets the requirements of the various experiments and the geology is shown in Figure 1. Excavation work should begin in early 1983 and should be terminated by the end of 1983. Most experiments would be started in 1984.

Most of the tunnelling will be drilled by full face tunnelling machine with a diameter of about 3.7 m. By this means, a smooth and relatively undisturbed surface will be obtained. The total length will be about 925 m. The infrastructure is concentrated in central area. The necessary cavern has dimensions of 13 x 30 x 9 m.

After branching off from the access tunnel, the tunnel system of the laboratory procedes parallel but with a lower gradient so that the existing test bore holes will remain in place above the laboratory.

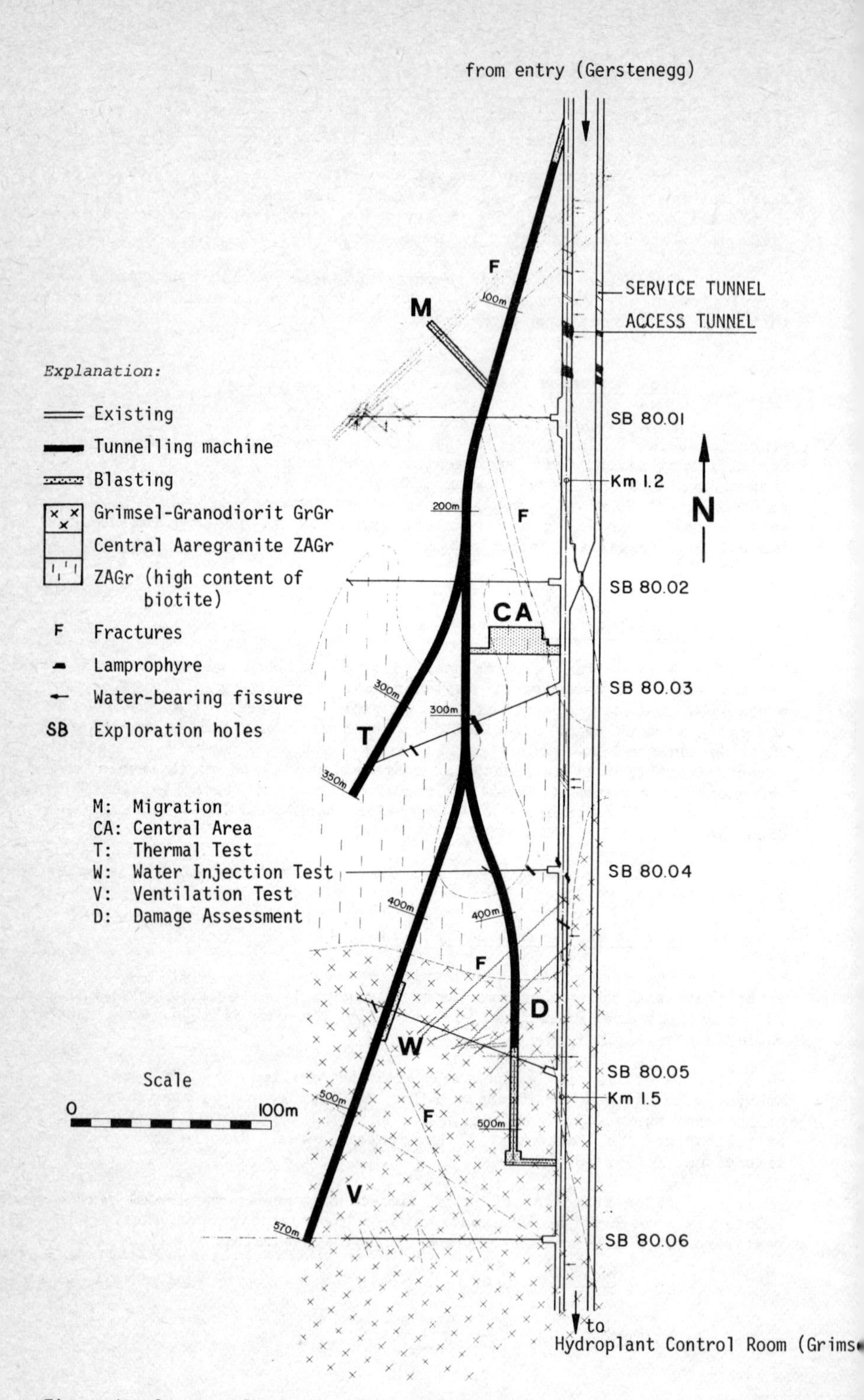

Figure 1. Conceptual Layout of the Grimsel Rock Laboratory

Geophysical Investigation

Designing, constructing and operating nuclear waste depositories requires a knowledge of the physical and hydraulic properties of the subsurface environment. In order to lower the risk of encountering unexpected site conditions, it is essential to have a powerful, nondestructive geophysical exploration tool. The currently available instrumentation is inadequate for meeting the goals of the nuclear waste program. The scale and objectives of petroleum and mineral exploration and exploration for engineering purposes are greatly different. The nuclear waste problem can require far greater precision and fineness of details.

The techniques and instrumentation to be developed should have the capability to determine the location, thickness and extent of fracture zones.

The most promising methods seem to be

- high frequency electromagnetic technique and
- seismic methods using low frequencies.

Electromagnetic prospecting using radiowave-frequencies has alredy been performed very successfully in salt mines to predict internal discontinuities. The boreholes with a minimum diameter of 28 mm had a maximum length of 3000 m and the measuring range was more than 100 m. These facts gave rise to the desire for implementing this method also for crystalline rock formations. Discontinuities can be located by employing high-frequency reflection or absorption techniques.

With the reflexion system, transmitter and receiver are situated in the same borehole whereas the absorption system allows the measurement of attenuation of radiofrequency-signals propagating between two boreholes. The reflexion system is based on the recording of ultra short pulses (40, 80 or 120 MHz) generated by a genuine transmitter arrangement. The disadvantage of the original reflexion probe is the lack of azimuthal resolution. Now, solutions which allow directional resolution are being investigated. The absorption system relies on generation and selective receiving of discrete frequencies (3 - 30 MHz without changing the mechanical length of the antenna), measuring attenuation in the propagating path with high accuracy.

A first series of measurements with the instrumentation used in the salt prospection was carried out in the exploration holes at Grimsel and showed good promise. Adjustment and development will be made of instrumentation, measurement procedures and interpretation techniques. The site at Grimsel will contain at least three parallel holes at an appropriate location which allows a partial excavation of the investigated rock in order to check the results obtained by the high frequency electromagnatic technique. The project using seismic methods with low frequencies is still in preparation.

Neotectonic Investigation

It could not be excluded that neotectonic movements of extensive blocks of rock still occur. The relative amount of movement, the extent of the moved block and the time history have a direct influence on the mechanical and geohydraulic behaviour of a waste repository with a lifetime of several 10'000 years. New flow paths may be opened or enlarged and others will be closed.

At Grimsel two approaches will be used to get an indication at the neotectonic movement:

- point measurements by high resolution tilt meters and
- line measurements by specially developed extensometers.

The tiltmeter which will be used at Grimsel has a usable measuring range of $\pm$ 25 $\cdot$ 10^{-6} rad and the resolution of the tiltmeter can be regarded as being better than 0.15 $\cdot$ 10^{-9} rad. By calibrating the tiltmeters frequently the changes in sensitivity over a long measuring period can be determined and the measurements can be adjusted accordingly. Tectonic and instrumental drifts may be separated by regularly rotating the instruments in the boreholes.

The measuring time at Grimsel will be at least one year. In order to get the complete deformation stage with line measurements by very sensitive extensometers it is necessary to measure in three prependiculer directions. The use of this measurement technique depends on the experience obtained with the tiltmeters.

Hydrogeological Investigations

The release of radionuclides to the biosphere from a repository is critically influenced by the hydrogeologic configuration and conditions of the site. The characterisation of the site requires the determination of critical piezometric conditions, permeability, effective porosity and storage coefficient. It is important to define and delineate the major dimensions of the fracture systems in terms of their spacing, orientation, aperture and continuity. This leads to qualitative understanding of whether the system behaves as a porous continuum or as a fractured discontinuum.

At the Grimsel Rock Laboratory the following projects will deal with the above mentioned problems:

- A modified water injection test
- A modified ventilation test
- A special set up for the measurement of the piezometric condition.

Modified Water Injection Test

The major objective of the experiment is to determine the directional hydraulic conductivity. A drift would be excavated to provide space for one or more ring-shaped arrays of boreholes. The direction of the holes will be more or less perpendicular to the most important water bearing joint system. Core analysis and tracer tests in a pilot hole will give an indication on the spacing and length of the boreholes. A tracer solution would be injected into various sections of the central hole while monitoring the hydraulic condition in selected intervals of the surrounding array. Hot salt water may be used as a tracer fluid.

The field tests on single fractures may lead to an indication of the hydraulically appropriate aperture and the area at the fracture. These data, together with the fracture density and orientation data from the core logging, can be used in computer programs for calculation of the hydraulic behaviour of the fracture network.

Modified Ventilation Test

At Stripa, Sweden, a macropermeability experiment has been carried out as part of the Swedish-American cooperative program (3). The experiment was an attempt to measure the average permeability of a very large valume of low-permeability, fractured rock. Inflow to the sealed 33 m long drift was measured by evaporating the seepage into the ventilation air while measuring the change in water vapor content between incoming and exhaust air streams. A similar test is currently under way at the Konrad mine in the Federal Republic of Germany.

Based upon the experience with these experiments, the test set up will be revised and an improved layout combined with a tracer test and geoelectrical measurements will be established.

Measurement of the Piezometric Conditions In order to understand the natural flow condition in fractured media it is necessary to mesure and monitor the hydraulic head at selected points in different boreholes. In Switzerland a new test set-up is currently being developed which allows continueous piezometric measurements along a borehole without taking any water. The instrument seems to be very suitable for long term observations.

Migration

To evaluate the potential of granite as a host rock for a nuclear waste repository, information is needed not only on fluid flow but also on ra ionuclide transport through fractures. In Switzerland research teams are developing computer codes to model the above mentioned processes and laboratory studies on sorption and radionuclide migration are being carried out at the Federal Institute of Reactor Rersearch. It is recognized worldwide that field tests are now needed to determine whether laboratory studies can accurately reflect in-situ conditions.

In the Grimsel laboratory preparatory work for tests designed to provide data for comparison of laboratory and field-measured retardation factors and to check the use of computer codes will be undertaken. The first task would be the sampling of uncontaminated water and the removal and transport of a fractured block of rock into the laboratory with minimum exposure of the fracture surface to the air.

In a next phase effort would be put on a migration test at a single fracture some what similar to those tests being carried out at Nevada Test Site (4) and at Stripa. In a third phase special emphasis would be put on studying channeling effects in a 3-D test.

Rock-Mass Response to Excavation

When an underground opening is excavated in crystalline rock some fracture characteristics are modified in the vicinity of the opening firstly due to mechanical destruction effects and secondly due to stress modifications. Blasting can modify the permeability in the immediate vicinity of the excavation by opening pre-existing fractures or by introduction of new fractures. Depending on the orientation of the stress tensor, stress modifications may alter the permeability in both diretions.

To study these effects, effort will be concentrated on measurements of density, hydraulic conductivity and deformation. Three different stages will be studied: undisturbed condition as a starting point, alteration during excavation and behaviour several time steps after the date of construction. Two test sections will be considered, one excavated by a tunneling machine and one by conventional blasting.

The physical arrangement for the proposed experiment consists of an upper and lower instrumentation drift which will be ready before excavating the experimental drift. Extensometers, stress meters and hydrogeologic testing and monitoring holes would be installed from these drifts and in arrays parallel to the experimental drift.

The density measurements would be made mainly by accoustic borehole logging in parallel, radial and longitudial boreholes spaced about 1 m apart.

The hydraulic conductivity measurements would be made by different types of water (air) injection tests. Tracer solutions may be used. In order to get a better understanding at the flow parallel to the drift, several slits perpendicular to the wall of the drift would be excavated. A central slit would be pressurized and the hydraulic condition of the neighbouring slits would be monitored. The deformation measurements would be taken mainly by special designed extensometers in the radial holes.

Improvement of Overcoring Technique

Three-dimensional stress measurements can be achieved today only by borehole deformation gauges combined with the use of overcoring techniques. However, 100 m is about the maximum depth commercially obtained with these techniques. Measurements at large depths can be performed at present only by hydraulic fracturing. This method cannot obtain the full stress tensor and requires coaxiality of the borehole and a principal stress direction.

The objective of the experiment is to improve several overcoring methods (borehole deformation gauges, biaxial and triaxial strain cells) for use at greater depth.

Thermal Behaviour of Rock Mass

High level waste represents a long lived heat source of decreasing output. In the Swiss concept the maximum temperature at the surface of the cannister should not exceed 100 °C.

Heat can have a great impact on the waste package itself (corrosion, solubility of the glass matrix etc.) and the backfill material between cannister and rock surface, especially if the buffer material is of some sort of clay (increase in creep rate, change in physio-chemical structure and chemical properties such as ion exchange capacity, etc.). These problems have been studied by several research teams in other in-situ investigations.

At Grimsel the thermal studies would concentrate on rock mass emphasis on

- convection currents and
- thermo-mechanical behaviour of joints.

The thermal load of the repository with its accompanying density gradients can cause thermal convective currents in the interstitial water and consequently shorter the transit times of particles from the repository to the biosphere. This is enhanced by the decrease in viscosity that water experiences as the temperature increases.

First computer-based analysis will be made in order to be decided if an in-situ test should be carried out at Grimsel.

Temperature changes result in thermally induced stresses which are super-imposed on the stresses induced by the excavation. In general, the higher stress levels decrease the overall stability, on the other hand the closure of fractures and increase in mean normal stress may improve mechanical characteristics and decrease permeability under certain conditions.

The test at Grimsel would be a single heater test with special emphasis on deformation and permeability studies of a single well defined fracture.

CONCLUSION

Some critical points of the storage of radioactive waste in crystalline formations have been addressed and a conceptual program of experiments is being developed. Detailed studies of each experiment will envolve the next year. Each experiment has to enhance the existing state of the art and should take advantage of the experience developing elsewhere.

REFERENCES

(1) Nagra, 1981: "Sondierbohrungen Juchlistock-Grimsel", Nagra Technical Report 81-07, November 1981

(2) Nagra, 1981: "Sondierbohrungen Juchlistock-Grimsel", Nagra Internal Report 81-01, Volume II, III, IV, V, September 1981

(3) P.A. Witherspoon, N.G.W. Cook, J.E. Gale: "Progress with Field Investigation at Stripa", LBL-10559, Februar 1980

(4) D. Isherwood, E. Raber, D. Coles, R. Stone: "Program Plan: Field Radionuclide Migration Studies in Climax Granite (Nevada Test Site) UCID-18838, November 1980

LIST OF PARTICIPANTS

LISTE DES PARTICIPANTS

BELGIUM - BELGIQUE

BUYENS, M., CEN/SCK, Boeretang 200, B-2400 Mol

NEERDAEL, B., CEN/SCK, Boeretang 200, B-2400 Mol

CANADA

DAVISON, C.C., Atomic Energy of Canada Limited, Whiteshell Nuclear Research Establishment, Pinawa, Manitoba ROE 1LO

FRITZ, P., Department of Earth Sciences, University of Waterloo, Waterloo, Ontario N2L 3GL

GALE, J.E., Memorial University of Newfoundland, St John's, Newfoundland

GRAY, M.N., Atomic Energy of Canada Limited, Whiteshell Nuclear Research Establishment, Pinawa, Manitoba ROE 1LO

LOPEZ, R.S., Atomic Energy of Canada Limited, Whiteshell Research Establishment, Pinawa, Manitoba ROE 1LO

PETERS, D.A., Atomic Energy of Canada Limited, Whiteshell Research Establishment, Pinawa, Manitoba ROE 1LO

SARGENT, F.P., Atomic Energy of Canada Limited, Whiteshell Research Establishment, Pinawa, Manitoba ROE 1LO

SIMMONS, G.R., Atomic Energy of Canada Limited, Whiteshell Research Establishment, Pinawa, Manitoba ROE 1LO

DENMARK - DANEMARK

ANDERSEN, L., Geological Survey of Denmark, Thoravej 31, DK-2400 Copenhagen NV

FINLAND - FINLANDE

ÄIKÄS, T., Saanio 1 Laine Oy, Mannerheimintie 31 A 3, SF-00250 Helsinki 25

ANTTILA, P., Imatran Voima Oy, P.O. Box 138, SF-00101 Helsinki

GARDEMEISTER, R., Imatran Voima Oy, P.O. Box 138, F-00101 Helsinki

HEINONEN, O.J., Technical Research Centre of Finland Reactor Laboratory, Otakaari 3 A, SF-02150 Espoo 15

HOLOPAINEN, A.-P., Technical Research Centre of Finland (VTT), Geotechnical Laboratory, Lämpömiehenkuja 2B, SF-02150 Espoo 15

JAKOBSSON, K.O., Institute of Radiation Protection, P.O. Box 268 SF-00101 Helsinki 10

KUUSI, J., Technical Research Centre of Finland Reactor Laboratory, Otakaari 3 A, F-02150 Espoo 15

LINDBERG, I.A., Geological Survey of Finland, Nuclear Waste Disposal Study Group, SF-02150 Espoo 15

MATIKAINEN, R.T., Helsinki University of Technology, Laboratory of Mining, 02150 Espoo 15

MIETTINEN, J., Department of Radiochemistry, University of Helsinki, Unioninkatu 35, SF-00170 Helsinki 17

PELTONEN, E.K., Technical Research Centre of Finland, Nuclear Engineering Laboratory, P.O.B. 169, SF-00181 Helsinki 18

PINNIGA, S., Department of Radiochemistry, University of Helsinki, Unioninkatu 35, SF-00170 Helsinki 17

RANTANEN, J., Technical Research Centre of Finland Reactor Laboratory, Otakaari 3 A, SF-02150 Espoo 15

RAUMOLIN, H., Industrial Power Company Ltd., Fredrikinkatu 51-53, SF-00100 Helsinki

RUUSKANEN, A., Imatran Voima Oy, P.O. Box 138, SF-00101 Helsinki

RYHÄNEN, V., Industrial Power Company Ltd., Fredrikinkatu 51-53, SF-00100 Helsinki

SALO, J.-P., Industrial Power Company Ltd., Fredrikinkatu 51-53, SF-00100 Helsinki

SÄRKKÄ, P.S., The Academy of Finland, Helsinki University of Technology, Dept Min. Metall., SF-02150 Espoo 15

TAHVANAINEN, T.A., Helsinki University of Technology, Laboratory of Mining Engineering, 02150 Espoo 15

VUORELA, P.K., Geological Survey of Finland, Nuclear Waste Disposal Study Group, SF-02150 Espoo 15

FRANCE

BARTHOUX, A., Adjoint au Directeur de l'ANDRA, Commissariat à l'Energie Atomique, 31-33, rue de la Fédération, 75015 Paris

DERLICH, S., Commissariat à l'Energie Atomique, IPSN/CSDR, B.P. N° 6, 92260 Fontenay-aux-Roses

SAAS, A., Commissariat à l'Energie Atomique, Agence Nationale pour la Gestion des Déchets Radioactifs (ANDRA), 31-33, rue de la Fédération, 75015 Paris

FEDERAL REPUBLIC OF GERMANY - REPUBLIQUE FEDERALE D'ALLEMAGNE

BREWITZ, W., Gesellschaft f. Strahlen.u. Umweltforschung m.b.H. München, Institut für Tieflagerung Wissenschaftliche, Abteilung Berliner Strasse 2, D-3392 Clausthal-Zellerfeld

FLACH, D., Gesellschaft f. Strahlen.u. Umweltforschung m.b.H. München, Institut für Tieflagerung Wissenschaftliche Abteilung, Berliner Strasse 2, D-3392 Clausthal-Zellerfeld

LIEDTKE, L., Bundesanstalt für Geowissenschaften und Rohstoffe (BGR), Postfach 510153, 3000 Hannover 51

ITALY - ITALIE

BRUZZI, D., ISMES, Viale Giulio Cesare 29, 24100 Bergama

JAPAN - JAPON

IIZUKA, Y., Shimizu Construction Co. Ltd., Civil Engineering Division Technology Department, N° 17-1, Kyobashi, 2-chome, Chuo-ku, Tokyo 104

KOJIMA, K., Department of Mineral Development, University of Tokyo, 7-3-1 Hongo, Bunkyo-ku, Tokyo 113

HAMADA, M., Civil Engineering Dept., Taisei Corporation, 25-1, Nishi-Shinjuku, 1-Chome, Shinjuku-ku, Tokyo

SAITO, S., Mitsubishi Metal Corporation (Nuclear Energy Department), N° 5-2, Ohte-machi, 1-chome, Chiyoda-ku, Tokyo 100

SEKIJIMA, K., Shimizu Construction Co. Ltd., Civil Engineering Division, Design Department, N° 17-1, Kyobashi, 2-chome, Chuo-ku, Tokyo 104

YAHIRO, T., Architectural Technology Development Dept., Kajima Corporation, 2-7, 1-chome, Motoakasaka, Minato-ku, Tokyo 107

NORWAY - NORVEGE

HUSEBY, S., Geological Survey of Norway, Drammensveien 230, Oslo 2

SWEDEN - SUEDE

ABELIN, M.H., Royal Institute of Technology, Roslagsvägen 101, S-104 05 Stockholm

AGESKOG, L., VBB/SWECO, Box 5038, S-102 41 Stockholm

AHAGEN, H., SKBF/KBS, Box 5864, 102 48 Stockholm

AHLBOM, K., SGAB, Box 1424, S-751 44 Uppsala

AHLSTRÖM, P.-E., SKBF/KBS, Swedish State Power Board, Fack, S-162 87 Vallingby

ANDERSSON, J.-E., SGAB, Box 1424, S-751 44 Uppsala

ANDERSSON, K.A., Swedish Nuclear Power Inspectorate, Box 27106, S-102 52 Stockholm

BERGMAN, S., Independant Consultant to SKBF/KBS, Föreningsvägen 19, S-182 74 Stocksund

BERGSTRÖM, A., SKBF/KBS, Box 5864, S-102 48 Stockholm

BÖRGESSON, L.E., University of Lulea, S-951 87 Lulea

BROTZEN, O., Yngvevägen 13, S-182 64 Djursholm

CARLSSON, H.S., Stripa Project Manager, SKBF/KBS, Box 5864, S-102 48 Stockholm

CARLSSON, L., SGAB SGU, Kungsgatan 4, S-411 19 Göteborg

EKLUND, S., Former Director General of IAEA, P.O. Box 100, A-1400 Vienna (Austria)

GIDLUND, J., Royal Institut of Technology, Dept. of Chemical Engineering, S-100 44 Stockholm

HALEN, P.A., Stripa Project, S-717 00 Stora

HANSBO, S., AB Jacobson & Widmark, Box 1214, S-181 23 Lidingo

HANSSON, K., SGAB, Box 1424, S-751 44 Uppsala

IRVINE, J.A., Stripa mine service, PL 8179 OLÖA, S-714 00 Kopparberg

KARLSSON, F.D.B., SKBF/KBS, Box 5864, S-102 48 Stockholm

KLOCKARS, C.-E., SGAB, Box 1424, S-751 44 Uppsala

LARSSON, A., Swedish Nuclear Power Inspectorate, Box 27106, S-102 52 Stockholm

LEIJON, B., Department of Rock Mechanics, University of Lulea, S-951 87 Lulea

NERETNIEKS, I., Royal Institute of Technology, Department of Chemical Engineering, S-100 44 Stockholm

NILSSON, J.E., AB Jaobson & Widmark, Box 199, S-951 23 Lulea

NILSSON, L.-B., SKBF/KBS, Box 5864, S-102 48 Stockholm

NORLANDER, H., Stripa Mine Service, S-717 00 Stora

OLSSON, N.T., K-Konsult, S-117 80 Stockholm

OLSSON, O.L., SGAB, Box 1424, S-751 44 Uppsala

PAPP, T., KBF/KBS, Box 5864, S-102 48 Stockholm

PUSCH, R., Div. of Soil Mechanics, University of Lulea, S-951 87 Lulea

RAMQVIST, G., Stripa Mine Service, S-717 00 Stora

RUNDQUVIST, G., Chairman, National Board for Spent Nuclear Fuel, Kungsgatan 35 III, S-111 56 Stockholm

RYDELL, N., National Board for Spent Nuclear Fuel, Kungsgatan 35 III, S-111 56 Stockholm

STEPHANSSON, O., Div. of Rock Mechanics, University of Lulea, S-951 87 Lulea

SVENKE, E., President of SKBF, SKBF, Box 5864, S-102 48 Stockholm

TEGELMARK, S.-E., Stripa Mine Service, S-717 00 Stora

THEGERSTRÖM, C., SKBF/KBS, Box 5864, S-102 48 Stockholm

SWITZERLAND - SUISSE

BECK, R.H., NAGRA, Nationale Genossenschaft für die Lagerung radioaktiver Abbfälle, Parkstrasse 23, CH-5400 Baden

EGGER, P., Rock Mechanics Lab., Federal Institute of Technology-Lausanne (EPFL), CH-1015 Lausanne

FRANK, E., Nuclear Safety Division, Swiss Federal Office of Energy, CH-5303 Würenlingen

FRITZ, P., Swiss Federal Institute of Technology, ISETH-Honggerberg, Zürich

GASSNER, R., NAGRA, Nationale Genossenschaft für die Lagerung radioaktiver Abbfälle, Parkstrasse 23, CH-5400 Baden

GRONEMEIER, K.U., NAGRA, Nationale Genossenschaft für die Lagerung radioaktiver Abbfälle, Parkstrasse 23, CH-5400 Baden

KAHR, G., c/o NAGRA, Parkstrasse 23, CH-5400 Baden

KEUSEN, H.R., Geotest AG, Birkenstrasse 15, CH-3052 Zollikofen

MÜLLER, W., Ingenieurunternehmung AG Bern, Thunstrasse 2, CH-3000 Bern 6

MÜLLER, W.H., NAGRA, Nationale Genossenschaft für die Lagerung radioaktiver Abbfälle, Parkstrasse 23, CH-5400 Baden

MÜLLER-VONMOOS, M., Swiss Federal Institute of Technology, CH-8092 Zürich

NOLD, A.L., NAGRA, Nationale Genossenschaft für die Lagerung radioaktiver Abbfälle, Parkstrasse 23, CH-5400 Baden

PFISTER, E., NAGRA, Nationale Genossenschaft für die Lagerung radioaktiver Abbfälle, Parkstrasse 23, CH-5400 Baden

THURY, M., NAGRA, Nationale Genossenschaft für die Lagerung radioaktiver Abbfälle, Parkstrasse 23, CH-5400 Baden

VAN DORP, F., NAGRA, Nationale Genossenschaft für die Lagerung radioaktiver Abbfälle, Parkstrasse 23, CH-5400 Baden

UNITED KINGDOM - ROYAUME-UNI

BLACK, J.H., Institute of Geological Sciences, Environmental Protection Unit, Building 151, Harwell Laboratory, Harwell, Oxfordshire OX11 0RA

HODGKINSON, D., U.K. Atomic Energy Authority, Theoretical Physics Div., Bldg. 424.4, Atomic Energy Res. Est., Harwell, Oxfordshire OX11 ORA

COOLING, C.M., Building Research Establishment, Department of the Environment, Geotechnics Division, Building Research Station Garston, Watford, Herts.

UNITED STATES - ETATS-UNIS

BALLARD, W., Office of Waste Isolation, U.S. Department of Energy, Washington D.C. 20545

BOYER, G., Office of Waste Isolation, U.S. Department of Energy, Washington D.C. 20545

GNIRK, P., Re/Spec., P.O. Box 725, Rapid City, SD 57701

HUSTRULID, W.A., Dept. of Mining, Colorado School of Mines, Golden, Colorado

MONTAZER, P., Colorado School of Mines, 31 Prospector Village, Golden, Colorado 80401

NORDSTROM, D.K., U.S. Geological Survey, MS-21, 345 Middlefield Road, Menlo Park, CA 94025

PATRICK, W.C., Lawrence Livermore National Laboratory, Spent Fuel Test-Climax, P.O. Box 808, Livermore, California 94550

ROBINSON, R., ONWI Project Management Division, 505 King Avenue, Columbus, Ohio 43201

WITHERSPOON, P.A., Associate Director, Lawrence Berkeley Laboratory, Earth Sciences Division, 90-1106 Berkeley, California 94720

INTERNATIONAL ATOMIC ENERGY AGENCY
AGENCE INTERNATIONALE DE L'ENERGIE ATOMIQUE

TSYPLENKOV, V., International Atomic Energy Agency, P.O. Box 100, A-1400 Vienna (Austria)

NUCLEAR ENERGY AGENCY - AGENCE POUR L'ENERGIE NUCLEAIRE

JOHNSTON, P.D., OECD Nuclear Energy Agency, Radiation Protection and Waste Management Division, 38 boulevard Suchet, 75016 Paris, France

SOME NEW PUBLICATIONS OF NEA

QUELQUES NOUVELLES PUBLICATIONS DE L'AEN

ACTIVITY REPORTS

RAPPORTS D'ACTIVITÉ

Activity Reports of the OECD Nuclear Energy Agency (NEA)

- 10th Activity Report (1981)
- 11th Activity Report (1982)

Rapports d'activité de l'Agence de l'OCDE pour l'Énergie Nucléaire (AEN)

- 10e Rapport d'Activité (1981)
- 11e Rapport d'Activité (1982)

Free on request – Gratuits sur demande

Annual Reports of the OECD HALDEN Reactor Project

- 21st Annual Report (1980)
- 22nd Annual Report (1981)

Rapports annuels du Projet OCDE de réacteur de HALDEN

- 21e Rapport annuel (1980)
- 22e Rapport annuel (1981)

Free on request – Gratuits sur demande

• • •

INFORMATION BROCHURES

- OECD Nuclear Energy Agency: Functions and Main Activities
- NEA at a Glance
- International Co-operation for Safe Nuclear Power
- The NEA Data Bank
- 25th Anniversary of the OECD Nuclear Energy Agency

BROCHURES D'INFORMATION

- Agence de l'OCDE pour l'Énergie Nucléaire : Rôle et principales activités
- Coup d'œil sur l'AEN
- Une coopération internationale pour une énergie nucléaire sûre
- La Banque de Données de l'AEN
- 25e Anniversaire de l'Agence de l'OCDE pour l'Énergie Nucléaire

Free on request – Gratuits sur demande

• • •

SCIENTIFIC AND TECHNICAL PUBLICATIONS

PUBLICATIONS SCIENTIFIQUES ET TECHNIQUES

RADIATION PROTECTION

RADIOPROTECTION

Recommendations for Ionization Chamber Smoke Detectors in Implementation of Radiation Protection Standards (1977)

Recommandations relatives aux détecteurs de fumée à chambre d'ionisation en application des normes de radioprotection (1977)

Free on request – Gratuit sur demande

Radon Monitoring
(Proceedings of the NEA Specialist Meeting, Paris, 1978)

Surveillance du radon
(Compte rendu d'une réunion de spécialistes de l'AEN, Paris, 1978)

£8.00 US$16.50 F66,00

Management, Stabilisation and Environmental Impact of Uranium Mill Tailings
(Proceedings of the Albuquerque Seminar, United States, 1978)

Gestion, stabilisation et incidence sur l'environnement des résidus de traitement de l'uranium
(Compte rendu du Séminaire d'Albuquerque, États-Unis, 1978)

£9.80 US$20.00 F80,00

Exposure to Radiation from the Natural Radioactivity in Building Materials
(Report by an NEA Group of Experts, 1979)

Exposition aux rayonnements due à la radioactivité naturelle des matériaux de construction
(Rapport établi par un Groupe d'experts de l'AEN, 1979)

Free on request – Gratuit sur demande

Marine Radioecology
(Proceedings of the Tokyo Seminar, 1979)

Radioécologie marine
(Compte rendu du Colloque de Tokyo, 1979)

£9.60 US$21.50 F86,00

Radiological Significance and Management of Tritium, Carbon-14, Krypton-85 and Iodine-129 arising from the Nuclear Fuel Cycle
(Report by an NEA Group of Experts, 1980)

Importance radiologique et gestion des radionucléides : tritium, carbone-14, krypton-85 et iode-129, produits au cours du cycle du combustible nucléaire
(Rapport établi par un Groupe d'experts de l'AEN, 1980)

£8.40 US$19.00 F76,00

The Environmental and Biological Behaviour of Plutonium and Some Other Transuranium Elements
(Report by an NEA Group of Experts, 1981)

Le comportement mésologique et biologique du plutonium et de certains autres éléments transuraniens
(Rapport établi par un Groupe d'experts de l'AEN, 1981)

£4.60 US$10.00 F46,00

Uranium Mill Tailing Management
(Proceedings of two Workshops)

La gestion des résidus de traitement de l'uranium
(Compte rendu de deux réunions de travail)

£7.20 US$16.00 F72,00

RADIOACTIVE WASTE MANAGEMENT

GESTION DES DÉCHETS RADIOACTIFS

Objectives, Concepts and Strategies for the Management of Radioactive Waste Arising from Nuclear Power Programmes
(Report by an NEA Group of Experts, 1977)

Objectifs, concepts et stratégies en matière de gestion des déchets radioactifs résultant des programmes nucléaires de puissance
(Rapport établi par un Groupe d'experts de l'AEN, 1977)

£8.50 US$17.50 F70,00

Decision of the OECD Council of the 22nd of July 1977 establishing a Multilateral Consultation and Surveillance Mechanism for Sea Dumping of Radioactive Waste

Décision du Conseil de l'OCDE en date du 22 juillet 1977 instituant un Mécanisme multilatéral de consultation et de surveillance pour l'immersion de déchets radioactifs en mer

Free on request – Gratuit sur demande

In Situ Heating Experiments in Geological Formations
(Proceedings of the Ludvika Seminar, Sweden, 1978)

Expériences de dégagement de chaleur in situ dans les formations géologiques
(Compte rendu du Séminaire de Ludvika, Suède, 1978)

£8.00 US$16.50 F66,00

Migration of Long-lived Radionuclides in the Geosphere
(Proceedings of the Brussels Workshop, 1979)

Migration des radionucléides à vie longue dans la géosphère
(Compte rendu de la réunion de travail de Bruxelles, 1979)

£8.30 US$17.00 F68,00

Low-Flow, Low-Permeability Measurements in Largely Impermeable Rocks
(Proceedings of the Paris Workshop, 1979)

Mesures des faibles écoulements et des faibles perméabilités dans des roches relativement imperméables
(Compte rendu de la réunion de travail de Paris, 1979)

£7.80 US$16.00 F64,00

On-Site Management of Power Reactor Wastes
(Proceedings of the Zurich Symposium, 1979)

Gestion des déchets en provenance des réacteurs de puissance sur le site de la centrale
(Compte rendu du Colloque de Zurich, 1979)

£11.00 US$22.50 F90,00

Recommended Operational Procedures for Sea Dumping of Radioactive Waste (1979)

Recommandations relatives aux procédures d'exécution des opérations d'immersion de déchets radioactifs en mer (1979)

Free on request – Gratuit sur demande

Guidelines for Sea Dumping Packages of Radioactive Waste
(Revised version, 1979)

Guide relatif aux conteneurs de déchets radioactifs destinés au rejet en mer
(Version révisée, 1979)

Free on request – Gratuit sur demande

Use of Argillaceous Materials for the Isolation of Radioactive Waste
(Proceedings of the Paris Workshop, 1979)

Utilisation des matériaux argileux pour l'isolement des déchets radioactifs
(Compte rendu de la réunion de travail de Paris, 1979)

£7.60 US$17.00 F68,00

Review of the Continued Suitability of the Dumping Site for Radioactive Waste in the North-East Atlantic (1980)

Réévaluation de la validité du site d'immersion de déchets radioactifs dans la région nord-est de l'Atlantique (1980)

Free on request – Gratuit sur demande

Decommissioning Requirements in the Design of Nuclear Facilities
(Proceedings of the NEA Specialist Meeting, Paris, 1980)

Déclassement des installations nucléaires : exigences à prendre en compte au stade de la conception
(Compte rendu d'une réunion de spécialistes de l'AEN, Paris, 1980))

£7.80 US$17.50 F70,00

Borehole and Shaft Plugging
(Proceedings of the Columbus Workshop, United States, 1980)

Colmatage des forages et des puits
(Compte rendu de la réunion de travail de Columbus, États-Unis, 1980)

£12.00 US$30.00 F120,00

Radionuclide Release Scenarios for Geologic Repositories
(Proceedings of the Paris Workshop, 1980)

Scénarios de libération des radionucléides à partir de dépôts situés dans les formations géologiques
(Compte rendu de la réunion de travail de Paris, 1980)

£6.00 US$15.00 F60,00

Research and Environmental Surveillance Programme Related to Sea Disposal of Radioactive Waste (1981)

Programme de recherches et de surveillance du milieu lié à l'immersion de déchets radioactifs en mer (1981)

Free on request – Gratuit sur demande

Cutting Techniques as related to Decommissioning of Nuclear Facilities
(Report by an NEA Group of Experts, 1981)

Techniques de découpe utilisées au cours du déclassement d'installations nucléaires
(Rapport établi par un Groupe d'experts de l'AEN, 1981)

£3.00 US$7.50 F30,00

Decontamination Methods as related to Decommissioning of Nuclear Facilities
(Report by an NEA Group of Experts, 1981)

Méthodes de décontamination relatives au déclassement des installations nucléaires
(Rapport établi par un Groupe d'experts de l'AEN, 1981)

£2.80 US$7.00 F28,00

Siting of Radioactive Waste Repositories in Geological Formations
(Proceedings of the Paris Workshop, 1981)

Choix des sites des dépôts de déchets radioactifs dans les formations géologiques
(Compte rendu d'une réunion de travail de Paris, 1981)

£6.80 US$15.00 F68,00

Near-Field Phenomena in Geologic Repositories for Radioactive Waste
(Proceedings of the Seattle Workshop, United States, 1981)

Phénomènes en champ proche des dépôts de déchets radioactifs en formations géologiques
(Compte rendu de la réunion de travail de Seattle, États-Unis, 1981)

£11.00 US$24.50 F110,00

Disposal of Radioactive Waste – An Overview of the Principles Involved, 1982

Évacuation des déchets radioactifs – Un aperçu des principes en vigueur, 1982

Free on request – Gratuit sur demande

Geological Disposal of Radioactive Waste – Research in the OECD Area (1982)

Évacuation des déchets radioactifs dans les formations géologiques – Recherches effectuées dans les pays de l'OCDE (1982)

Free on request – Gratuit sur demande

Geological Disposal of Radioactive Waste: Geochemical Processes (1982)

Évacution des déchets radioactifs dans les formations géologiques : Processus géochimique (1982)

£7.00 US$14.00 F70,00

Geological Disposal of Radioactive Waste: In Situ experiments in Granite (Proceedings of the Stockholm Workshop, 1982)

Évacuation des déchets radioactifs dans les formations géologiques : Expériences in situ dans du granite (Compte rendu d'une réunion de travail de Stockholm, 1982)

£10.00 US$20.00 F100,00

Interim Oceanographic description of the North-East Atlantic Site for the Disposal of Low-Level Radioactive Waste (1983)

État des connaissances océanographiques relatives au site d'immersion de déchets radioactifs de faible activité dans l'Atlantique Nord-Est (1983)

Free on request – Gratuit sur demande

The International Stripa Project: Background and Research Results (1983)

Projet international de Stripa : Informations générales et résultats des recherches (1983)

Free on request – Gratuit sur demande

• • •

SCIENTIFIC INFORMATION

INFORMATION SCIENTIFIQUE

Calculation of 3-Dimensional Rating Distributions in Operating Reactors (Proceedings of the Paris Specialists' Meeting, 1979)

Calcul des distributions tridimensionnelles de densité de puissance dans les réacteurs en cours d'exploitation (Compte rendu de la Réunion de spécialistes de Paris, 1979)

£9.60 US$21.50 F86,00

Nuclear Data and Benchmarks for Reactor Shielding (Proceedings of a Specialists' Meeting, 1979)

Données nucléaires et expériences repères en matière de protection des réacteurs (Compte rendu d'une réunion de spécialistes, Paris, 1980)

£9.60 US$24.00 F96,00

• • •

NUCLEAR FUEL CYCLE

LE CYCLE DU COMBUSTIBLE NUCLÉAIRE

World Uranium Potential – An International Evaluation (1978)

Potentiel mondial en uranium – Une évaluation internationale (1978)

£7.80 US$16.00 F64.00

Uranium – Resources, Production and Demand (1982)

Uranium – Ressources, production et demande (1982)

£9.90 US$22.00 F99,00

Nuclear Energy and Its Fuel Cycle: Prospects to 2025

L'énergie nucléaire et son cycle de combustible : perspectives jusqu'en 2025

£11.00 US$24.00 F110,00

Dry Storage of Spent Fuel Elements (Proceedings of an NEA Specialist Workshop, Madrid, 1982).

Stockage à sec des éléments combustibles irradiés (Compte rendu d'une réunion de spécialistes de l'AEN, Madrid, 1982).

£8.50 US$17.00 F85,00

Uranium Exploration Methods – Review of the NEA/IAEA R & D Programme (Proceedings of the Paris Symposium, 1982)

Les méthodes de prospection de l'uranium – Examen du programme AEN/AIEA de R & D (Compte rendu du symposium de Paris, 1982).

£24.00 US$48.00 F240,00

Uranium Extraction Technology – Current practice and new developments on ore processing (A NEA/IAEA Joint Report, 1982)

Les techniques d'extraction de l'uranium – Pratiques actuelles et nouveautés dans le domaine du traitement du minerai (Rapport conjoint AEN/AIEA, 1982)

£10.00 US$20.00 F100,00

R & D Uranium Exploration Techniques – Joint OECD-NEA/IAEA Newsletter (on regular basis)

R & D sur les techniques de prospection de l'uranium – Bulletin d'information conjoint OCDE-AEN/AIEA (série)

Free on request – Gratuit sur demande

R & D in Uranium Extraction Technology – Joint OECD-NEA/IAEA Newsletter (once a year)

R & D sur les techniques d'extraction de l'uranium – Bulletin d'information conjoint OCDE-AEN/AIEA (1 numéro par an)

Free on request – Gratuit sur demande

NUCLEAR ENERGY PROSPECTS

PERSPECTIVES DE L'ÉNERGIE NUCLÉAIRE

Nuclear Energy Prospects to 2000 (A joint Report by NEA/IEA)

Perspectives de l'Énergie Nucléaire jusqu'en 2000 (Rapport conjoint AEN/AIE)

£7.00 US$14.00 F70,00

SAFETY

Nuclear Aerosols in Reactor Safety
(A State-of-the-Art Report by a Group of Experts, 1979)

Plate Inspection Programme
(Report from the Plate Inspection Steering Committee – PISC – on the Ultrasonic Examination of Three Test Plates), 1980

Reference Seismic Ground Motions in Nuclear Safety Assessments
(A State-of-the-Art Report by a Group of Experts, 1980)

Nuclear Safety Research in the OECD Area. The Response to the Three Mile Island Accident (1980)

Safety Aspects of Fuel Behaviour in Off-Normal and Accident Conditions
(Proceedings of the Specialist Meeting, Espoo, Finland, 1980)

Safety of the Nuclear Fuel Cycle
(A State-of-the-Art Report by a Group of Experts, 1981)

Critical Flow Modelling in Nuclear Safety
(A State-of-the-Art Report by a Group of Experts, 1982)

SÛRETÉ

Les aérosols nucléaires dans la sûreté des réacteurs
(Rapport sur l'état des connaissances établi par un Groupe d'experts, 1979)

£8.30 US$18.75 F75,00

Programme d'inspection des tôles
(Rapport du Comité de Direction sur l'inspection des tôles – PISC – sur l'examen par ultrasons de trois tôles d'essai au moyen de la procédure «PISC» basée sur le code ASME XI), 1980

£3.30 US$7.50 F30,00

Les mouvements sismiques de référence du sol dans l'évaluation de la sûreté des installations nucléaires
(Rapport sur l'état des connaissances établi par un Groupe d'experts, 1980)

£7.00 US$16.00 F64,00

Les recherches en matière de sûreté nucléaire dans les pays de l'OCDE. L'adaptation des programmes à la suite de l'accident de Three Mile Island (1980)

£3.20 US$8.00 F32,00

Considérations de sûreté relatives au comportement du combustible dans des conditions anormales et accidentelles
(Compte rendu de la réunium de spécialistes, Espoo, Finlande, 1980)

£12.60 US$28.00 F126,00

Sûreté du Cycle du Combustible Nucléaire
(Rapport sur l'état des connaissances établi par un Groupe d'experts, 1981)

£6.60 US$16.50 F66,00

La modélisation du débit critique et la sûreté nucléaire
(Rapport sur l'état des connaissances établi par un Groupe d'experts, 1982)

£6.60 US$13.00 F66,00

• • •

LEGAL PUBLICATIONS

PUBLICATIONS JURIDIQUES

Convention on Third Party Liability in the Field of Nuclear Energy – incorporating the provisions of Additional Protocol of January 1964

Convention sur la responsabilité civile dans le domaine de l'énergie nucléaire – Texte incluant les dispositions du Protocole additionnel de janvier 1964

Free on request – Gratuit sur demande

Nuclear Legislation, Analytical Study: "Nuclear Third Party Liability" (revised version, 1976)

Législations nucléaires, étude analytique : « Responsabilité civile nucléaire » (version révisée, 1976)

£6.00 US$12.50 F50.00

Nuclear Legislation, Analytical Study: "Regulations governing the Transport of Radioactive Materials" (1980)

Législations nucléaires, étude analytique : « Réglementation relative au transport des matières radioactives » (1980)

£8.40 US$21.00 F84,00

Nuclear Legislation, Analytical Study: Regulatory and Institutional Framework for Nuclear Activities (1983)
(in preparation)

Législations nucléaires, étude analytique : Réglementation générale et cadre institutionnel des activités nucléaires (1983)
(en préparation)

£00 US$00 F00

Nuclear Law Bulletin
(Annual Subscription – two issues and supplements)

Bulletin de Droit Nucléaire
(Abonnement annuel – deux numéros et suppléments)

£6.00 US$13.00 F60,00

Index of the first thirty issues of the Nuclear Law Buletin

Index des trente premiers numéros du Bulletin de Droit Nucléaire

Description of Licensing Systems and Inspection of Nuclear Installation (1980)

Description du régime d'autorisation et d'inspection des installations nucléaires (1980)

£7.60 US$19.00 F76,00

NEA Statute

Statuts de l'AEN

Free on request – Gratuit sur demande

OECD SALES AGENTS
DÉPOSITAIRES DES PUBLICATIONS DE L'OCDE

ARGENTINA – ARGENTINE
Carlos Hirsch S.R.L., Florida 165, 4° Piso (Galería Guemes)
1333 BUENOS AIRES, Tel. 33.1787.2391 y 30.7122
AUSTRALIA – AUSTRALIE
Australia and New Zealand Book Company Pty, Ltd.,
10 Aquatic Drive, Frenchs Forest, N.S.W. 2086
P.O. Box 459, BROOKVALE, N.S.W. 2100
AUSTRIA – AUTRICHE
OECD Publications and Information Center
4 Simrockstrasse 5300 BONN. Tel. (0228) 21.60.45
Local Agent/Agent local :
Gerold and Co., Graben 31, WIEN 1. Tel. 52.22.35
BELGIUM – BELGIQUE
CCLS – LCLS
19, rue Plantin, 1070 BRUXELLES. Tel. 02.521.04.73
BRAZIL – BRÉSIL
Mestre Jou S.A., Rua Guaipa 518,
Caixa Postal 24090, 05089 SAO PAULO 10. Tel. 261.1920
Rua Senador Dantas 19 s/205-6, RIO DE JANEIRO GB.
Tel. 232.07.32
CANADA
Renouf Publishing Company Limited,
2182 St. Catherine Street West,
MONTRÉAL, Que. H3H 1M7. Tel. (514)937.3519
OTTAWA, Ont. K1P 5A6, 61 Sparks Street
DENMARK – DANEMARK
Munksgaard Export and Subscription Service
35, Nørre Søgade
DK 1370 KØBENHAVN K. Tel. +45.1.12.85.70
FINLAND – FINLANDE
Akateeminen Kirjakauppa
Keskuskatu 1, 00100 HELSINKI 10. Tel. 65.11.22
FRANCE
Bureau des Publications de l'OCDE,
2 rue André-Pascal, 75775 PARIS CEDEX 16. Tel. (1) 524.81.67
Principal correspondant :
13602 AIX-EN-PROVENCE : Librairie de l'Université.
Tel. 26.18.08
GERMANY – ALLEMAGNE
OECD Publications and Information Center
4 Simrockstrasse 5300 BONN Tel. (0228) 21.60.45
GREECE – GRÈCE
Librairie Kauffmann, 28 rue du Stade,
ATHÈNES 132. Tel. 322.21.60
HONG-KONG
Government Information Services,
Publications/Sales Section, Baskerville House,
2/F., 22 Ice House Street
ICELAND – ISLANDE
Snaebjörn Jönsson and Co., h.f.,
Hafnarstraeti 4 and 9, P.O.B. 1131, REYKJAVIK.
Tel. 13133/14281/11936
INDIA – INDE
Oxford Book and Stationery Co. :
NEW DELHI-1, Scindia House. Tel. 45896
CALCUTTA 700016, 17 Park Street. Tel. 240832
INDONESIA – INDONÉSIE
PDIN-LIPI, P.O. Box 3065/JKT., JAKARTA, Tel. 583467
IRELAND – IRLANDE
TDC Publishers – Library Suppliers
12 North Frederick Street, DUBLIN 1 Tel. 744835-749677
ITALY – ITALIE
Libreria Commissionaria Sansoni :
Via Lamarmora 45, 50121 FIRENZE. Tel. 579751/584468
Via Bartolini 29, 20155 MILANO. Tel. 365083
Sub-depositari :
Ugo Tassi
Via A. Farnese 28, 00192 ROMA. Tel. 310590
Editrice e Libreria Herder,
Piazza Montecitorio 120, 00186 ROMA. Tel. 6794628
Costantino Ercolano, Via Generale Orsini 46, 80132 NAPOLI. Tel. 405210
Libreria Hoepli, Via Hoepli 5, 20121 MILANO. Tel. 865446
Libreria Scientifica, Dott. Lucio de Biasio "Aeiou"
Via Meravigli 16, 20123 MILANO Tel. 807679
Libreria Zanichelli
Piazza Galvani 1/A, 40124 Bologna Tel. 237389
Libreria Lattes, Via Garibaldi 3, 10122 TORINO. Tel. 519274
La diffusione delle edizioni OCSE è inoltre assicurata dalle migliori librerie nelle città più importanti.
JAPAN – JAPON
OECD Publications and Information Center,
Landic Akasaka Bldg., 2-3-4 Akasaka,
Minato-ku, TOKYO 107 Tel. 586.2016
KOREA – CORÉE
Pan Korea Book Corporation,
P.O. Box n° 101 Kwangwhamun, SÉOUL. Tel. 72.7369
LEBANON – LIBAN
Documenta Scientifica/Redico,
Edison Building, Bliss Street, P.O. Box 5641, BEIRUT.
Tel. 354429 – 344425
MALAYSIA – MALAISIE
and/et **SINGAPORE - SINGAPOUR**
University of Malaya Co-operative Bookshop Ltd.
P.O. Box 1127, Jalan Pantai Baru
KUALA LUMPUR. Tel. 51425, 54058, 54361
THE NETHERLANDS – PAYS-BAS
Staatsuitgeverij
Verzendboekhandel Chr. Plantijnstraat 1
Postbus 20014
2500 EA S-GRAVENHAGE. Tel. nr. 070.789911
Voor bestellingen: Tel. 070.789208
NEW ZEALAND – NOUVELLE-ZÉLANDE
Publications Section,
Government Printing Office Bookshops:
AUCKLAND: Retail Bookshop: 25 Rutland Street,
Mail Orders: 85 Beach Road, Private Bag C.P.O.
HAMILTON: Retail Ward Street,
Mail Orders, P.O. Box 857
WELLINGTON: Retail: Mulgrave Street (Head Office),
Cubacade World Trade Centre
Mail Orders: Private Bag
CHRISTCHURCH: Retail: 159 Hereford Street,
Mail Orders: Private Bag
DUNEDIN: Retail: Princes Street
Mail Order: P.O. Box 1104
NORWAY – NORVÈGE
J.G. TANUM A/S Karl Johansgate 43
P.O. Box 1177 Sentrum OSLO 1. Tel. (02) 80.12.60
PAKISTAN
Mirza Book Agency, 65 Shahrah Quaid-E-Azam, LAHORE 3.
Tel. 66839
PHILIPPINES
National Book Store, Inc.
Library Services Division, P.O. Box 1934, MANILA.
Tel. Nos. 49.43.06 to 09, 40.53.45, 49.45.12
PORTUGAL
Livraria Portugal, Rua do Carmo 70-74,
1117 LISBOA CODEX. Tel. 360582/3
SPAIN – ESPAGNE
Mundi-Prensa Libros, S.A.
Castelló 37, Apartado 1223, MADRID-1. Tel. 275.46.55
Libreria Bosch, Ronda Universidad 11, BARCELONA 7.
Tel. 317.53.08, 317.53.58
SWEDEN – SUÈDE
AB CE Fritzes Kungl Hovbokhandel,
Box 16 356, S 103 27 STH, Regeringsgatan 12,
DS STOCKHOLM. Tel. 08/23.89.00
SWITZERLAND – SUISSE
OECD Publications and Information Center
4 Simrockstrasse 5300 BONN. Tel. (0228) 21.60.45
Local Agents/Agents locaux
Librairie Payot, 6 rue Grenus, 1211 GENÈVE 11. Tel. 022.31.89.50
TAIWAN – FORMOSE
Good Faith Worldwide Int'l Co., Ltd.
9th floor, No. 118, Sec. 2
Chung Hsiao E. Road
TAIPEI. Tel. 391.7396/391.7397
THAILAND – THAILANDE
Suksit Siam Co., Ltd., 1715 Rama IV Rd,
Samyan, BANGKOK 5. Tel. 2511630
TURKEY – TURQUIE
Kültur Yayinlari Is-Türk Ltd. Sti.
Atatürk Bulvari No : 77/B
KIZILAY/ANKARA. Tel. 17 02 66
Dolmabahce Cad. No : 29
BESIKTAS/ISTANBUL. Tel. 60 71 88
UNITED KINGDOM – ROYAUME-UNI
H.M. Stationery Office, P.O.B. 569,
LONDON SE1 9NH. Tel. 01.928.6977, Ext. 410 or
49 High Holborn, LONDON WC1V 6 HB (personal callers)
Branches at: EDINBURGH, BIRMINGHAM, BRISTOL,
MANCHESTER, BELFAST.
UNITED STATES OF AMERICA – ÉTATS-UNIS
OECD Publications and Information Center, Suite 1207,
1750 Pennsylvania Ave., N.W. WASHINGTON, D.C.20006 – 4582
Tel. (202) 724.1857
VENEZUELA
Libreria del Este, Avda. F. Miranda 52, Edificio Galipan,
CARACAS 106. Tel. 32.23.01/33.26.04/33.24.73
YUGOSLAVIA – YOUGOSLAVIE
Jugoslovenska Knjiga, Terazije 27, P.O.B. 36, BEOGRAD.
Tel. 621.992

Les commandes provenant de pays où l'OCDE n'a pas encore désigné de dépositaire peuvent être adressées à :
OCDE, Bureau des Publications, 2, rue André-Pascal, 75775 PARIS CEDEX 16.

Orders and inquiries from countries where sales agents have not yet been appointed may be sent to:
OECD, Publications Office, 2 rue André-Pascal, 75775 PARIS CEDEX 16.

65958-12-1982

PUBLICATIONS DE L'OCDE, 2, rue André-Pascal, 75775 PARIS CEDEX 16 - N° 42526 1983
IMPRIMÉ EN FRANCE
(66 83 02 3) ISBN 92-64-02416-6